2015 Home Builders' Jobsite Codes:
A Quick Guide to the 2015 International Residential Code

Stephen A. Van Note

T0206762

NAHB BuilderBooks

INTERNATIONAL
CODE COUNCIL®

2015 Home Builders' Jobsite Codes:
A Quick Guide to the 2015 International Residential Code

BuilderBooks, a Service of the National Association of Home Builders

Elizabeth M. R. Hartke	Acquisitions & Managing Editor
Circle Graphics	Cover Design
Electronic Quill Publishing Services	Composition
King Printing Company	Printing
Gerald M. Howard	NAHB Chief Executive Officer
Lakisha A. Woods, CAE	NAHB Senior Vice President & Chief Marketing Officer
Mark A. Johnson	ICC Executive Vice President, Business Development
Suzane Olmos	ICC Director of Products and Special Sales
Steve Van Note, CBO	ICC Managing Director, Product Development
Hamid Naderi, P.E., CBO	ICC Senior Vice President, Product Development

Disclaimer

This publication is provided for information purposes only. It is not intended to serve as a substitute for the *2015 International Residential Code,* as a basis for complying with it, or as an alternative to any state or local building code or ordinance. It is not offered for the purpose of providing professional advice, either from a legal, accounting, or other professional service standpoint. It is recommended that you consult with an experienced professional in the interpretation and/or application of the information so provided. Reference herein to any specific commercial products, process, or service by trade name, trademark, manufacturer, or otherwise does not necessarily constitute or imply its endorsement, recommendation, or favored status by the National Association of Home Builders. The views and opinions of the author expressed in this publication do not necessarily state or reflect those of the National Association of Home Builders, and they shall not be used to advertise or endorse a product.

Printed in the United States of America

20 19 18 17   2 3 4 5

ISBN-13: 978-0-86718-741-0
eISBN-13: 978-0-86718-742-7

For further information, please contact:

National Association of Home Builders	International Code Council
1201 15th Street, NW	500 New Jersey Avenue, NW, 6th Floor
Washington, DC 20005-2800	Washington, DC 20001-2070
800-223-2665	888-ICC-SAFE (422-7233)
BuilderBooks.com	iccsafe.org

About the Author

Stephen A. Van Note is managing director of product development for the International Code Council (ICC), where he is responsible for developing technical resource materials in support of the International Codes. He also manages the review and technical editing of staff-written publications as well as those written by external authors. In addition, Van Note develops and presents *International Residential Code* seminars nationally. Prior to joining ICC in 2006, Van Note was a building official for Linn County, Iowa. He has 15 years of experience in code administration and enforcement, and more than 20 years of experience in the construction field, including project planning and management for residential, commercial, and industrial buildings. A certified building official and plans examiner, Van Note also holds certifications in several inspection categories.

Contents

Figures

Tables

Chapter 5. Floors

Chapter 6. Wall Construction

Chapter 7. Wall Covering

Chapter 8. Roof Framing

Acknowledgments

The author is grateful to Ron Van Note, master carpenter and builder, for his usual expert counsel.

Introduction

Home Builders' Jobsite Codes is a field guide for
builders, trade contractors, design professionals,
inspectors, and others involved in the design and
construction of residential buildings. It is based on
the *2015 International Residential Code®* (IRC). This
comprehensive stand-alone code establishes minimum regulations for the construction of one- and
two-family *dwellings* and *townhomes*. It includes provisions for

- Structural design
- Fire and life safety
- Energy conservation
- Mechanical systems
- Fuel-gas systems
- Plumbing
- Electrical systems

The IRC's purpose is to safeguard public safety,
health, and general welfare from fire and other

potential hazards attributed to the built environment, while keeping homes affordable. The code
provides for strong, stable, and sanitary homes that
conserve energy and yet offer adequate lighting and
ventilation.

Published by the International Code Council®
(ICC), the IRC is maintained and updated through
an open code development process and is available internationally for adoption by governing
authorities.

Home Builders' Jobsite Codes focuses on the *prescriptive* provisions of the IRC—"recipes," if you
will, for meeting code requirements without the
need for an engineered design. These provisions
address all aspects of conventional construction of
dwellings and their accessory buildings. Illustra-
tions and tables assist the reader in understanding
the code requirements and address frequently asked
questions. Some terms appear in italics the first time
they are used in the text. The glossary at the back of
the book defines these terms. *Home Builders' Jobsite
Codes* also includes other useful information not in
the IRC, such as weights of building materials and
components.

Although this guide is organized into chapters
similar to the IRC, there are some important exceptions. For example, for ease of use, IRC "Building

Planning" requirements, detailed in chapter 3 of the code itself, are divided into three separate chapters in this pocket guide:
1. Structural Design Criteria
2. Fire Safety
3. Safe and Healthy Living Environments

Home Builders' Jobsite Codes explores these important issues in more detail in other chapters as well.

Home Builder's Jobsite Codes is not an official code, and has not been adopted as such in any jurisdiction. The publication intends to serve as a guide only. It does not include all applicable requirements of the IRC. For example, certain *performance* criteria related to engineered design are outside the limited scope of this publication. Builders should consult the IRC, local amendments, and local building departments for more detailed requirements and for criteria related to other methods of construction.

Builders also should obtain specific information on design criteria for wind, snow, seismic (earthquake) events, flood, soil, or other atmospheric and geological conditions, as well as any amendments to the code, from their local building departments. Since code requirements for energy conservation, decay resistance, and termite control requirements also may vary by geographic region, builders should obtain that information from local building code departments.

For more information on the *2015 International Residential Code,* go online to the International Code Council's Web site, www.iccsafe.org, or call 1-888-ICC-SAFE (422-7233).

1

Structural Design Criteria

The IRC establishes minimum structural design criteria necessary to accommodate normal loads placed on a building and, depending on a home's location, resist the forces of natural hazards such as snow, wind, earthquake, and flood. In most cases, the tried-and-true construction practices offered in the IRC incorporate these criteria, eliminating the need for an engineered design or complex calculations. For example, the code provides span tables for conventional wood framing elements such as joists, girders, headers, and rafters.

Construction must safely support all loads:

- Snow, wind, seismic, and flood loads, which vary by geographic region
- *Live loads*
- *Dead loads*
- Roof loads

Note: *The roof is designed for the roof live load (not more than 20 psf) or the snow load, whichever is greater.*

To correctly apply the values of the tables and the prescriptive methods of construction, builders must know the structural design criteria in the planning chapter of the code. Determining the appropriate live loads is fairly straightforward. However, seismic, wind, snow, soil, or flood area values differ by geographic location. In addition, frost depth, weathering severity, ice barrier underlayment requirements, and history of termite damage vary by climate and geography. Therefore, builders often must obtain information through the maps found in the IRC or through their local building departments.

Moreover, some structural elements still may require an engineered design. For example, the sizing of wide-flange steel beams commonly used in dwelling construction is outside the scope of the IRC. Instead, accepted engineering practices will determine their sizes.

Live Loads

Minimum required live loads for floors are based on the use of the space. Guards and handrails also must be secured to safely resist forces against them (table 1.1).

Table 1.1 Minimum uniformly distributed live loads

Use	Live load (psf)	Note
Rooms other than sleeping rooms	40	
Sleeping rooms	30	
Decks and exterior balconies	40	
Stairs	40	Concentrated load of 300 lb. per 4 sq. in.
Habitable attics and attics served by fixed stairs	30	
Uninhabitable attics with limited storage	20	Access hatch or pull-down stair to storage area at least 24 in. wide × 42 in. high
Uninhabitable attics without storage	10	
Passenger vehicle garages	50	Elevated garage floors must support a concentrated load of 2,000 lb. per 20 sq. in.
Handrails and top rails of guards		Concentrated load of 200 lb. applied from any direction
Guard balusters and infill panels		Horizontally applied load of 50 lb. on an area of 1 sq. ft.

Deflection

Allowable deflection is a measurement of bending under code-prescribed loads to ensure adequate stiffness of structural framing members such as studs, joists, beams, and rafters (table 1.2). Although the prescriptive tables account for deflection in their values, builders must be familiar with deflection limits in order to choose the appropriate table for sizing a framing member. Allowable deflection is measured by dividing the span or length (L) of the member by a prescribed factor, such as 360 for floor joists (L/360). To determine allowable deflection for a certain span, convert feet to inches and divide the result

Table 1.2 Allowable deflection of structural members

Structural member	Allowable deflection
Rafters, slope > $\frac{3}{12}$, no finished ceiling attached to rafters	L/180
Rafters, slope > $\frac{3}{12}$, finished ceiling attached to rafters	L/240
Ceiling joist	L/240
Plastered ceilings	L/360
Floors	L/360
All other structural members	L/240

Note: Wall deflection and wind load deflections are not shown.

by 360. The following example is for a floor joist with a 16 ft. span:

L = 16 ft. × 12 in. = 192 in.

Allowable deflection = 192 in. / 360 = 0.53 in.

Allowable deflection for this floor joist is approximately ½ in.

Note: A 16 ft. span rafter with a ⁴⁄₁₂ slope and no ceiling attached has an allowable deflection of L/180, which is twice the deflection allowed for floor joists.

Calculating Dead Loads

The prescriptive tables of the IRC detailing continuous footing sizes for conventional frame construction assume average weights of construction materials. Therefore, additional calculations typically are not required. The material and component weights (tables 1.3 and 1.4) may help builders correctly size an isolated footing, or another element not covered in the IRC tables.

Wind

The prescriptive structural provisions of the IRC are limited to those geographical regions with ultimate design wind speeds of 140 mph or less (130 mph in hurricane-prone regions) as defined in the IRC

Table 1.3 Building material weights

Materials	Weight (psf)
Plywood – ¼ in.	.8
Plywood – ½ in.	1.6
Plywood – ¾ in.	2.4
4" Brick	35.0
Gypsum board – ½ in.	2.1
Gypsum board – ⅝ in.	2.5
Quarry tile – ½ in.	5.8
Hardwood flooring – ²⁵⁄₃₂ in.	4.0
Built-up roofing	6.5
Shingles, asphalt	1.7–2.8
Shingles, wood	2.0–3.0
Common dimension lumber (lb. per cu. ft.)	27–29 lb. per cu. ft.
Concrete (lb. per cu. yd.)	150 lb. per cu. ft.

wind maps. Otherwise, the code requires a design in accordance with one of the referenced standards. In addition to an engineered design that complies with the *International Building Code (IBC)*[1] and ASCE 7,[2] the IRC includes references to ICC 600, *Standard for Residential Construction in High Wind Regions*[3] and *AWC Wood Frame Construction Manual (WFCM).*[4]

Table 1.4 Average weights of building components

Description	Weight (psf)
Roof dead load (framing, sheathing, asphalt shingles, insulation, drywall)	10
Exterior wall (2 × 4 framing, sheathing, siding, insulation, drywall)	10
Floor (joist, sheathing, carpeting, drywall)	10
Concrete wall–8 in. thick	100
10 in. thick	125
12 in. thick	150
Concrete block wall–8 in. thick	60

Wind Exposure Category

In addition to the basic wind speeds for a geographic area, ground surface irregularities affect the wind's impact on a building. The IRC classifies wind exposure into three categories:

1. Exposure B—some wind protection with trees and buildings characteristic of urban and suburban settings
2. Exposure C—open terrain with scattered obstructions
3. Exposure D—adjacent to large bodies of water, including hurricane-prone regions

Exposure categories are important design criteria for engineering purposes. For many of the prescriptive methods of wood frame construction in the IRC, wind exposure category is not a factor. However, wind exposure category must be considered when applying the provisions for wall sheathing, wood wall bracing, roof tie-down, and exterior wall and roof coverings. The following components must be *listed* and installed to resist wind loads based on the wind speed and exposure category:

- Siding
- Roof covering
- Windows
- Skylights
- Exterior doors
- Overhead doors

Hurricanes

Hurricane-prone regions are the coastal areas of the Atlantic Ocean and Gulf of Mexico where the ultimate design wind speed is greater than 115 mph. The IRC wind maps identify the portions of hurricane-prone regions that require an engineered design or a design that complies with other referenced standards. Windows and other glazing require additional protection if they are in *windborne debris regions*—those areas within hurricane-prone regions as specifically defined in the code.

Storm Shelters

Storm shelters, sometimes called *safe rooms,* are not required by the code. However, they offer added protection from the destructive forces of high winds, hurricanes, and tornadoes. When installed within a dwelling or as a separate structure, storm shelters must conform to the requirements of ICC-500, *Standard on the Design and Construction of Storm Shelters.*[5]

Earthquake

The IRC assigns a *seismic design category (SDC)* to building sites relative to the anticipated intensity and frequency of earthquakes. (For more details, see the *seismic* map in the code.) For buildings located in SDC A or B and constructed under the prescriptive methods of the IRC, there are no additional seismic requirements. One- and two-family dwellings in SDC C also are exempt from the seismic requirements. However, specific seismic requirements apply to townhomes sited in SDC C, and to all buildings in SDC D_0, D_1, and D_2.

The higher seismic design categories (SDC D_0, D_1, and D_2) occur predominantly in western parts of the U.S., in the New Madrid area of Missouri, Illinois, Indiana, Tennessee, and Arkansas, and in South Carolina.

2

Fire Safety

Fire-resistant construction is required to separate *dwelling units,* for exterior walls located close to property lines, and on the garage side of an attached garage/dwelling separation.

This chapter addresses these requirements, as well as protection of floor assemblies and foam plastic insulation, fire sprinkler systems, smoke alarms, and emergency escape. Fire protection related to specific building components is addressed throughout the code and is detailed in this and other chapters of this book.

Location on Property

Exterior walls of dwellings and garages must maintain a minimum separation from the property line, measured perpendicular to the wall, or be protected for fire. For dwellings without fire sprinkler systems, walls less than 5 ft. from the property line must have a fire-resistance rating of one hour when exposed to fire from either side of the wall. The code permits a 3 ft. separation without a fire-resistance-rated wall when the dwelling has an automatic fire sprinkler system. A typical one-hour wall consists of ⅝ in. Type X gypsum board or gypsum sheathing on both sides of a frame wall with insulation filling the cavity. However, a number of other approved designs are available.

The following fire-resistance requirements are based on distances to the property line (figs. 2.1 and 2.2):

For dwellings without fire sprinklers:

- Less than 5 ft.—one-hour wall
- Less than 5 ft.—limited area for windows and doors
- Less than 3 ft.—no windows or doors
- Less than 2 ft.—no roof overhang projections

Roof overhang projections less than 5 ft. from the property line require one-hour protection on their underside.

For dwellings with fire sprinklers:

- Less than 3 ft.—one-hour wall
- Less than 3 ft.—no windows or doors
- Less than 2 ft.—no roof overhang projections

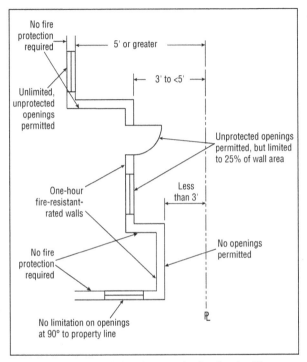

Figure 2.1 Exterior wall location near lot line for dwellings without fire sprinklers

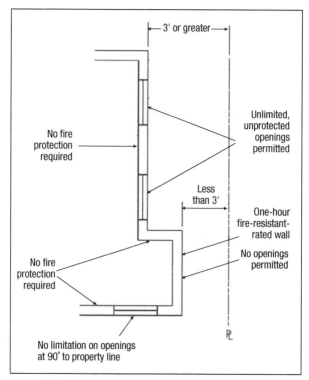

Figure 2.2 Exterior wall location near lot line for dwellings with fire sprinklers

Roof overhang projections less than 3 ft. from the property line require one-hour protection on their undersides.

Note: Sheds and playhouses of 200 sq. ft. or less do not require fire-resistant protection regardless of their location on the property.

Dwelling Unit Separation

Fire-resistant-rated separations are required between dwelling units of duplexes and townhomes as follows:

Two-Family Dwelling Separation

- One-hour fire-resistance-rated construction is required to completely separate dwelling units.
- The separation may have a ½-hour rating if an NFPA 13 sprinkler system is installed throughout.[6]

Note: NFPA 13R[7] and 13D[8] systems do not satisfy this requirement.

- The separating wall assemblies must extend from the foundation to the underside of the roof sheathing and to the exterior walls (fig. 2.3). There is one exception:
  - Wall assemblies need not extend through the *attic* if ⅝ in. Type X gypsum board is installed

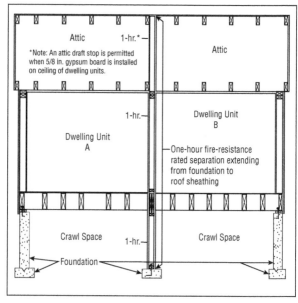

Figure 2.3 Fire-resistant-rated separation between dwelling units of duplex

on the ceiling of the living space and an attic draft stop lines up with the separating wall.

Townhome Separation

The fire-resistance rating of the common wall separating townhome dwelling units depends on the presence or absence of an automatic fire sprinkler system. Townhomes without sprinkler protection require a 2-hr. rated separation whereas the rating is reduced to 1 hr. for townhomes with sprinkler systems (fig. 2.4). Additional requirements follow:

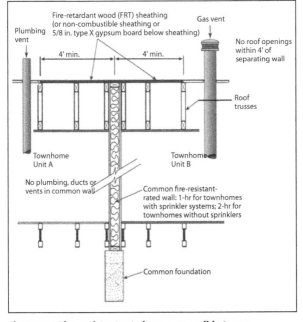

Figure 2.4 Fire-resistant-rated common wall between townhomes

- The common separation wall cannot contain plumbing, mechanical equipment, vents, or ducts.
- The separation wall must be continuous from the foundation to the underside of the roof sheathing.
- Parapets are not required when certain fire-resistant protection is provided for the roof for 4 ft. on both sides of the separating wall.

Fire-Resistant-Rated Penetrations

When items such as pipes or ducts penetrate one or both sides of the fire-resistant-rated wall or the floor/ceiling assemblies separating dwelling units, both the penetrating item and the space around it must be protected to maintain fire resistance. In general, penetrations by steel, ferrous and copper pipe, tubing, and conduit require only that the space left around the pipe be filled with approved materials to limit the passage of fire. The material, such as a listed fire caulk, must be installed according to the manufacturer's instructions to provide a fire-resistant time rating equivalent to that of the construction being penetrated.

Other penetrating materials, such as plastic pipe or metal ducts, (not permitted for the common wall separating townhomes), must be protected further by approved means, such as a listed firestop assembly or a fire damper, to maintain the fire-resistance rating of the wall or floor/ceiling assembly. Steel electrical boxes up to 16 sq. in. and listed electrical boxes of any material are permitted if they are separated by approved means when they are located on opposite sides of the wall. Wood blocking or applying listed putty pads to both boxes are two means of separating them.

Garages

Attached garages and detached garages within 3 ft. of the residence require protection to resist the spread of fire to the dwelling, but not the same protection as required between dwelling units (fig. 2.5). The separation does not require an assembly with a fire-resistance rating as in the case of dwelling unit separations. The prescriptive requirements are as follows:

- To provide complete separation of the garage from the dwelling, ½ in. gypsum board must be installed on the garage side.
- If habitable rooms are located above the garage, ⅝ in. Type X gypsum board must be installed on the garage ceiling and ½ in. gypsum board must be installed on the walls supporting the floor/ceiling.

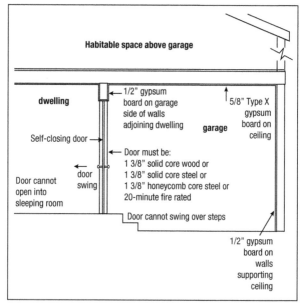

Figure 2.5 Garage separation from dwelling (habitable space above garage)

- Doors between the dwelling and garage must be self-closing and one of the following types:
  - 1⅜ in.-thick solid wood
  - 1⅜ in.-thick solid-core steel
  - 1⅜ in.-thick honeycomb-core steel
  - 20-minute fire rated

- Ducts in the garage as well as ducts penetrating the walls or ceilings that separate the dwelling from the garage shall be of a minimum No. 26 gage sheet steel. The ducts shall have no openings into the garage.
- Garages cannot open into a sleeping room.
- Garage floors shall be noncombustible and sloped to a drain or the vehicle entry door.

Fire Protection of Floors

For certain types of floor framing such as open-web trusses and I-joists, the IRC requires application of a membrane to the underside of the framing to provide limited fire protection and delay collapse of the floor in a fire. The membrane must be one of the following materials:

- ½ in. gypsum wallboard
- ⅝ in. wood structural panel
- Other equivalent material
   Protection is not required for floor assemblies:
- Over a space protected with automatic fire sprinklers
- Over a crawl space not intended for storage or fuel-fired appliances
- With unprotected portions not greater than 80 sq. ft. per story isolated by prescribed fireblocking

▪ Of 2 × 10 or greater dimension lumber or structural composite lumber

Foam Plastic Insulation

A thermal barrier of ½ in. gypsum board or equivalent must separate foam plastic from the interior. The following exceptions are permitted:

▪ In attics and crawl spaces entered only for repairs or maintenance, foam plastic may be covered with one of the following ignition-barrier materials:
- 1½ in. mineral fiber insulation
- ¼ in. *wood structural panels*
- ⅜ in. particleboard
- ¼ in. hardboard
- ⅜ in. gypsum board
- 0.016 in. corrosion-resistant steel
- 1½-inch-thick cellulose insulation
- ¼-inch fiber-cement panel, soffit or backer board

▪ Foam plastic may be spray applied to a sill plate and header (rim joist area) without the thermal barrier when the foam plastic meets the following requirements:
- No more than 3¼ in. thick
- Density of 1½–2 lb. per cu. ft.
- Flame-spread index less than or equal to 25 and smoke-developed index less than or equal to 450

Other types of insulation and insulation facings or vapor barriers also may require covering with ½ in. gypsum board or other approved materials, depending on flame-spread and smoke-developed properties of the exposed material.

Automatic Fire Sprinkler Systems

A dwelling unit fire sprinkler system helps to detect and control fires and is meant to prevent *flashover* in the room where a fire originates to allow time for occupants to escape a building. The code requires an automatic fire sprinkler system in new townhomes and new one- and two-family dwellings.

The sprinkler system must conform to one of the following:

- IRC sec. P2904, Dwelling Unit Fire Sprinkler Systems

    or

- NFPA 13D, Standard for the Installation of Sprinkler Systems in One- and Two-Family Dwellings and Manufactured Homes

A dwelling unit fire sprinkler system requires less water compared with NFPA 13 and 13R systems. It may be either a multipurpose or stand-alone system. The code permits connection to many types of water supply systems to meet the required fire sprinkler system capacity. Generally, a dwelling unit fire sprinkler system does not require sprinklers in the following locations:

- Closets not greater than 24 sq. ft. with a depth of 3 ft. or less
- *Bathrooms* no larger than 55 sq. ft.
- Open porches
- Garages
- Attics without a fuel-fired appliance
- Crawl spaces
- Concealed spaces

Dwelling fire sprinkler systems also do not require a fire department connection.

Note: *The IRC requires a single sprinkler above fuel-fired equipment installed in an attic.*

Smoke Alarms

Dwelling units require smoke alarms as follows:

Required Locations

In new homes and dwelling units undergoing remodeling, the code requires a smoke alarm in the following locations:

- Each sleeping room
- Outside of and in the immediate vicinity of each separate sleeping area
- On each story, including basements and *habitable attics*

Required Clearances

To prevent nuisance alarms, the code requires the following minimum clearances for smoke alarms:

- 3 ft. horizontally from a bathroom door
- 20 ft. horizontally from a cooking appliance for ionization smoke alarms
- 10 ft. horizontally from a cooking appliance for ionization smoke alarms with an alarm-silencing switch
- 6 ft. horizontally from a cooking appliance for photoelectric smoke alarms

Installation Requirements

The smoke alarms must meet the following installation requirements:

- Hardwired to the building power with no disconnection switch
- Battery backup
- Alarms interconnected

Note: *The code allows listed wireless alarms in place of physical interconnection. All alarms must sound upon activation of one alarm.*

Additions, Alterations, and Repairs

For construction that requires a permit, other than deck and porch additions and exterior work, such as siding, windows, doors, and roofing, dwelling units must have smoke alarms located as required for new construction.

Smoke alarms may be battery operated without connection to the building wiring. The code does not require interconnection of alarms under the following conditions:

- When alterations do not remove interior wall and ceiling finishes
- If there is no attic or under-floor space that provides access to run new wiring

Emergency Escape and Rescue Openings

One of the most important safety provisions in the IRC is for emergency escape and rescue openings. These openings provide a secondary means to exit a sleeping room or basement if a fire occurs. They must allow occupants to escape directly to a yard or open space and allow rescue personnel fully equipped with breathing apparatus to enter the room from the outside. Typically these openings are windows or doors; however, other openings may be approved.

Required Locations

Occupants are most vulnerable to the hazards of fire when they are not fully alert or when they are in spaces that have no windows or doors. The code addresses these life-safety issues by requiring an emergency escape and rescue opening in the following locations:

- Every sleeping room
- Habitable attics
- Basements, unless they are not more than 200 sq. ft. and are used only for mechanical equipment or a storm shelter

Note: *For other than new sleeping rooms, the code does not require installation of an emergency escape and rescue opening in an existing basement undergoing remodeling or repairs.*

General Requirements

In an emergency, occupants need to move quickly and easily to the safety of the outdoors. Therefore, emergency escape and rescue openings must

- Open to a public way, yard, or court
- Open from the inside without the use of keys, tools, or special knowledge

Dimensions

In order for emergency escape and rescue openings to serve their intended purpose, the code limits the sill height above the floor and prescribes minimum dimensions as follows (table 2.1 and fig. 2.6):

- 5.7 sq. ft. net opening (5 sq. ft. for grade floor or below grade openings)
- 24 in. minimum net opening height

Table 2.1 Minimum net clear width/height combinations for 5.7 sq. ft. emergency escape and rescue openings (in.)

Width	20	20.5	21	21.5	22	22.5
Height	41	40	39.1	38.2	37.3	36.5
Width	23	23.5	24	24.5	25	25.5
Height	35.7	34.9	34.2	33.5	32.8	32.2
Width	26	26.5	27	27.5	28	28.5
Height	31.6	31	30.4	29.8	29.3	28.8
Width	29	29.5	30	30.5	31	31.5
Height	28.3	27.8	27.4	26.9	26.5	26.1
Width	32	32.5	33	33.5	34	34.2
Height	25.7	25.3	24.9	24.5	24.1	24

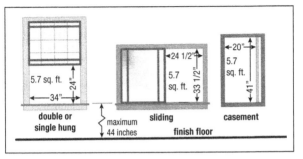

Figure 2.6 Emergency escape and rescue windows

- 20 in. minimum net opening width
- 44 in. maximum sill height
- 36-in.-high path to yard or court when the emergency escape window is under a deck or porch

Window Well
The code requires a window well if the sill of the emergency escape opening is below grade (fig. 2.7). Window wells shall

- Be at least 9 sq. ft. in area
- Have a minimum dimension of 36 in.
- Allow the emergency escape element to fully open

If a window well is more than 44 in. deep, it must have a ladder or steps. Ladders may encroach

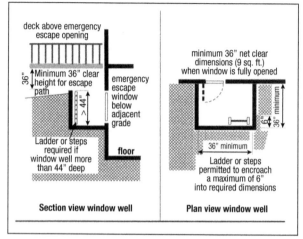

Figure 2.7 Window well for emergency escape and rescue opening

6 in. into the required space. They must meet the following requirements:

- Be at least 12 in. wide
- Project at least 3 in.
- Have rungs spaced not more than 18 in. apart

Approved bars, grilles, or covers are permitted over emergency escape windows or window wells if they

▪ Maintain net opening dimensions
▪ Can be released or removed from the inside without the use of a key, tool, special knowledge, or force greater than that required for normal operation of the escape and rescue opening

Window wells serving emergency escape openings must have provisions for drainage to the foundation drain.

3

Safe and Healthy Living Environments

Although structural integrity and fire protection
were addressed generally in the previous chapters,
this chapter includes other minimum requirements
for constructing safe and healthy living environments, including *means of egress,* carbon monoxide
alarms, room dimensions, ventilation, protection
from decay, and safety glazing.

Means of Egress

Stairways, ramps, hallways, and doors are the
primary means of egress from dwelling units.
The means of egress provides a continuous unobstructed path from all portions of the dwelling to
the outdoors through the required exit door without passing through the garage. There must be at
least one exterior exit door meeting the minimum
size requirements. Otherwise, the code does not

specify minimum door sizes. Hallways must be at least 3 ft. wide. To provide fire protection for the egress path, enclosed accessible space under stairs must be protected on the enclosed side with ½ in. gypsum board.

Doors

All dwelling units require at least one exterior exit door that meets the following criteria:

- Side hinged
- At least 3 ft. wide (32 in. net opening)
- At least 6 ft. 8 in. high (78 in. net opening)
- Readily openable from the inside without the use of a key or special knowledge or effort

Landings for Doors

Generally, each exterior door must have a landing on the exterior side that is at least as wide as the door and 36 in. in the direction of travel (fig. 3.1). At the required exit door, the maximum threshold height is

- 1½ in. above the interior floor
- 7¾ in. above the exterior landing when the door swings in
- 1½ in. above the exterior landing when the door swings out

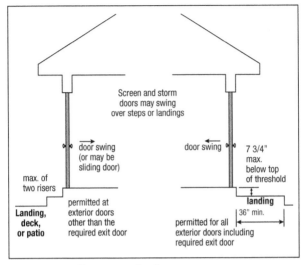

Figure 3.1 Landings at exterior doors

For exterior doors other than the required exit door, the maximum threshold height is

- 7¾ in. above the interior floor
- 7¾ in. above the exterior landing

The exterior side of a door does not require a landing under the following conditions:

- The door is not the required exit door.

- The stairway on the exterior side has two or fewer risers.
- The door, other than an exterior storm or screen door, does not swing over the stairway.

Note: The maximum slope for exterior landings is ¼:12 (2%).

Note: The code does not require a 36 in. landing for exterior balconies that are smaller than 60 sq. ft. that are only accessible from the inside.

Stairways

The code addresses stair safety through provisions for proper pitch, walking surface, clearances, uniformity, and graspable handrails.

Stairway Width and Headroom

As part of the egress path, the IRC establishes minimum dimensions for width and headroom height for a safe stairway. The minimum clear width is greater above the handrails to accommodate a person's shoulder width. Headroom is measured vertically above the nose of the tread. The dimension requirements are as follows:

- Clear width above the handrail is at least 36 in.
- Handrail projection is not more than 4½ in. on each side.

- ■ The minimum clear width at and below the handrail is
  - 31½ in. if there is handrail on one side
  - 27 in. if there is handrail on both sides
- ■ Headroom must be at least 6 ft. 8 in.

Stair Treads and Risers

The minimum tread depth and maximum riser height determine the maximum stair steepness but just as important for stair safety is the uniformity of treads and risers for the full *flight* of the stairs (fig. 3.2). The requirements are

- ■ A maximum riser height of no more than 7¾ in.
- ■ No more than ⅜ in. tolerance between the largest and smallest riser.
- ■ A minimum tread depth of 10 in.
- ■ No more than ⅜ in. tolerance between the largest and the smallest tread.
- ■ Winder treads must be at least 10 in. deep at 1 ft. in from the narrow side and not less than 6 in. at any point.

Stair Profile

In most cases, the IRC requires solid risers and tread nosings that project in front of the solid riser. The nosing dimensions are:

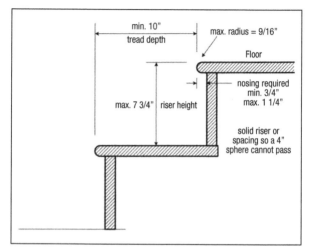

Figure 3.2 Stair tread and riser detail

- Projection of ¾ to 1¼ in. beyond riser
- Radius not greater than ⁹⁄₁₆ in.

Nosings are not required when the tread depth is 11 in. or greater.

Solid risers are not required if one of the following criteria is met:

- The opening between treads does not allow the passage of a 4 in. sphere.

- The opening is not more than 30 in. above the floor or grade.

Landings at Stairways

A floor or landing typically is required at the top and bottom of stairs.

- A landing is not required at the top of interior stairs if a door does not swing over stairs. (This requirement applies to attached garages and other interior stairs.)
- Landings must be at least the width of the stair and 36 in. in the direction of travel.
- No more than a 147 in. rise is permitted between landings.

Handrails

A handrail is required on at least one side of each stair flight with four or more risers. The handrail must comply with these requirements (fig. 3.3):

- Height 34–38 in. above the nose of the tread
- Ends returned to a wall or terminated in newel posts or safety terminals
- Space of at least 1½ in. between the wall and the handrail
- Continuous for the full flight, from a point above the bottom riser to the top riser

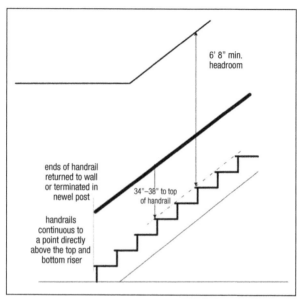

Figure 3.3 Stair and handrail

- Handrails may be interrupted by a newel post at a turn.
- Volute, turnout, starting easing, or starting newel may be used over the lowest tread.

Approved Handrail Grip Sizes
The size and shape of handrails must be suitable for grasping to prevent falls on stairs. The code offers

three options for satisfying this requirement as follows (fig. 3.4):

- ▦ Circular with a 1¼–2 in. diameter
- ▦ Noncircular, perimeter of 4–6¼ in. with a cross section of no more than 2¼ in.
- ▦ Noncircular, perimeter more than 6¼ in. with a graspable finger recess area on both sides of the profile
  - Finger-recess depth of at least 5⁄16 in.
  - Width of the handrail above the recess, 1¼–2¾ in.

Note: *Other shapes that provide equivalent graspability are also acceptable.*

Spiral Stairs

Spiral stairways are an exception to the general stair requirements and are permitted, as follows, with no limitation on the area or *habitable space* they serve (fig. 3.5).

- ▦ Width at least 26 in.
- ▦ At least 6¾ in. tread depth at 12 in. from the narrower edge
- ▦ All treads identical
- ▦ Risers no more than 9½ in.
- ▦ Headroom at least 6 ft. 6 in.

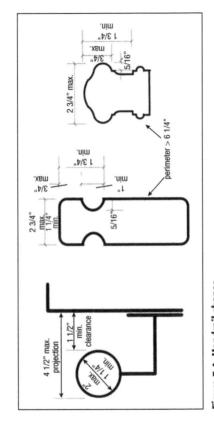

Figure 3.4 Handrail shapes

46

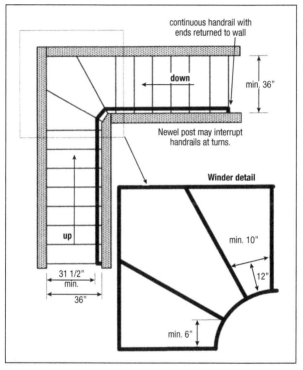

Figure 3.5 Winder treads

Ramps

As is the case for stairs, handrails, guards, and other elements, the IRC governs only ramps that are a part of the dwelling, accessory building, or structure. Separate elements such as landscaping features, including stairs, ramps, sidewalks, and driveways are not regulated by the code. The code limits the slope of ramps as follows:

- Maximum 1:8 (12.5%) slope for ramps that do not serve the required egress door
- Maximum 1:12 (8.3%) slope for ramps serving the required egress door
  Note: Ramps serving the required egress door may have a slope of up to 1:8 (12.5%) if site constraints prohibit a 1:12 slope.
- 3 × 3 ft. landings at the following locations:
  - Top and bottom of ramps
  - Where doors open onto ramps
  - Where ramps change direction
- Handrails on at least one side of all ramps exceeding a slope of 1:12 (8.3% slope).

Guards

To protect against falls and injury, the code requires guards at the edges of elevated walking surfaces that are more than 30 in. above the floor or ground at any point within 36 in. of the edge of the walking surface (figs. 3.6 and 3.7).

Guard Height

When the code requires a guard, the minimum guard height is

- 36 in. above the walking surface of decks, balconies, porches, landings, floors, and ramps
- 34 in. above the tread nosing on the open sides of stairs

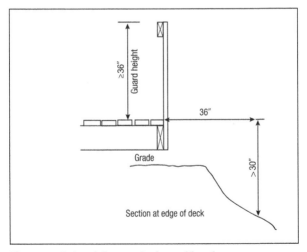

Figure 3.6 Determining required guard locations

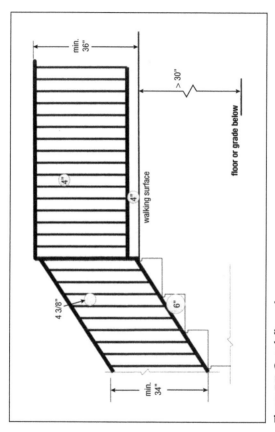

Figure 3.7 Guard dimensions

50

Guard Openings

The code limits the size of any openings in balusters or other infill components of required guards to prevent passage of the following:

- 4 in. diameter sphere at most locations
- 4⅜ in. diameter sphere at the sides of stairs
- 6 in. diameter sphere at the triangle formed by the riser, tread, and bottom rail of a guard at the open side of a stairway

Window Sill Height

To reduce injuries to children from falls through windows, the code requires the lowest part of the clear opening of the window to be at least 24 in. above the finished floor if the window opening is more than 72 in. above the grade or surface below (fig. 3.8). If the opening is less than 24 in. in these locations, installation can include a barrier or limits on the window dimensions as follows:

- Openings in operable portions of windows that do not allow passage of a 4-in.-diameter sphere
- Approved window opening control device
- Approved window fall-prevention device

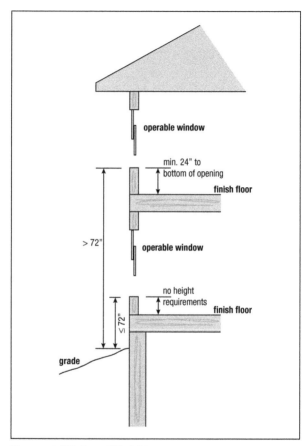

Figure 3.8 Window sill height

Carbon Monoxide Alarms

The code requires carbon monoxide alarms as follows:

Required Locations

In new homes and dwelling units undergoing remodeling, the code requires a carbon monoxide alarm outside each separate sleeping area.

Note: An alarm is also required in a bedroom if the bedroom or attached bathroom contains a fuel-burning appliance.

Note: Dwelling units without an attached garage and with no fuel-fired appliances do not require carbon monoxide alarms.

Installation Requirements

The carbon monoxide alarms must meet the following installation requirements:

- Connected to the building power with no disconnection switch
- Battery backup

Additions, Alterations, and Repairs

For construction that requires a permit, other than deck and porch additions and exterior work, such as siding, windows, doors, and roofing, dwelling units must have carbon monoxide alarms located as required for new construction. However, the carbon monoxide alarms may be battery operated without connection to the building wiring.

Light, Ventilation, Heating

The code sets minimum requirements for natural or artificial light, fresh air ventilation, and heating in cold climates, to create a healthy and livable environment.

Light

In almost all cases, electric lighting satisfies the minimum illumination requirement of 6 *footcandles* for habitable rooms and bathrooms. However, windows may still be required for ventilation and emergency escape and rescue. Interior and exterior stairs require artificial lighting to illuminate the landings and treads. The light source must be located

- To illuminate interior landings and treads to a minimum level of 1 foot-candle
- In the immediate vicinity of the top landing at exterior locations
- At the bottom landing of exterior stairs leading to a basement

For lighting that is not continuous or automatic, the code requires control by a switch located at each floor level for interior stairs with six or more risers

Ventilation

Homes that comply with the energy efficiency provisions of the code are of sufficiently tight construction to require mechanical ventilation to bring in fresh air. Although openable windows are still desirable for a healthy living environment, the IRC requires a *whole-house mechanical ventilation system* when the air infiltration rate is within the limits set by the energy requirements in the dwelling unit as determined by a blower door test. As part of this system, bathrooms typically have *local exhaust* (an exhaust fan) and openable windows are optional.

Outside air intake openings must be

- At least 10 ft. from any noxious source, such as plumbing vents or chimneys
- Protected on the outside with corrosion-resistant screens with an opening of ¼ in. to ½ in.

Heating

Dwellings must incorporate heating where the winter design temperature, as determined by the local jurisdiction, is less than 60°F. All habitable rooms must be capable of maintaining a minimum temperature of 68°F at 3 ft. above the floor and 2 ft. from exterior walls.

Minimum Room Dimensions

Minimum dimensions and areas for rooms and hallways are as follows:
- Habitable rooms except kitchens, at least 70 sq. ft., with a minimum dimension of 7 ft.
- Hallway width, at least 3 ft.

Minimum ceiling height measured from the finished floor is 7 ft. except as follows:
- Sloped ceiling, 7 ft. for 50% of required area
- Bathrooms and laundry rooms, 6 ft. 8 in.
- In showers, 6 ft. 8 in.
- Basements with no habitable space, 6 ft. 8 in.
  Note: Beams and ducts may be 6 ft. 4 in. from the floor in basements.

Sanitation

Minimum sanitary requirements for each dwelling unit are as follows:
- Connection to approved water supply and approved sewer or private septic system
- One water closet, lavatory, and bathtub or shower

- Kitchen with sink
- Hot and cold water to sinks, lavatories, laundry, bathtubs, and showers

Toilet, Bath, and Shower Spaces

Showers must have nonabsorbent wall surfaces to 6 ft. above the floor. They must be at least 30 × 30 in. with 24 in. in front of opening. Other minimum clearances for bathroom fixtures are as follows (fig. 3.9):

- 21 in. in front of water closet and lavatory
- 15 in. from centerline of water closet or lavatory to wall, vanity cabinet, or tub
- 30 in. from centerline of lavatory to centerline of water closet

    Note: The code permits a narrow shower space of 25 in. or more when the area of the shower is not less than 1,300 sq. in. (for example, 25 in. × 52 in.)

Safety Glazing

When glazing is installed in hazardous locations where human impact could occur, the code requires safety glazing, such as tempered or laminated glass. In most cases, the glazing must be etched or permanently *labeled* to identify the type of glazing and the safety glazing standard with which it complies. The code defines the following locations with glazing

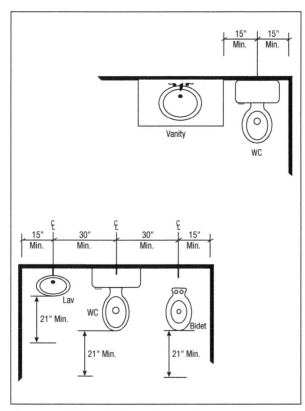

Figure 3.9 Minimum bathroom fixture clearances

as hazardous and, therefore, requires safety glazing (fig. 3.10):

▪ Fixed and operable panels of all swinging, sliding, and bifold doors

▪ Within 24 in. of the operable panel of a door in a closed position and where the glazing's bottom edge is less than 60 in. above the floor, unless

  • There is an intervening wall.

  • The door is to a closet no more than 3 ft. deep.

  • The wall with glazing is perpendicular to the closed door and the door does not swing toward the glazing.

▪ Walls, enclosures, or fences containing or facing hot tubs, spas, whirlpools, bathtubs, showers, and swimming pools where the bottom exposed edge of the glazing is less than 60 in. above any standing or walking surface unless the glazing is more than 60 in. from the water's edge of a bathtub, hot tub, spa, whirlpool, swimming pool, or from the edge of a shower.

▪ Wherever there is an individual pane larger than 9 sq. ft. with its bottom edge less than 18 in. above the floor, and its top edge more than 36 in. above the floor

▪ Less than 36 in. above the walking surface and adjacent to stairways, ramps and intermediate landings between flights (fig. 3.11)

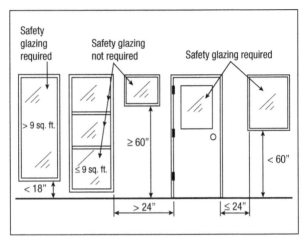

Figure 3.10 Safety glazing

■ Less than 36 in. above the walking surface of the bottom landing and within 60 in. horizontally of the bottom tread of a stairway

Mirrors and other glass panels mounted or hung on a surface that provides a continuous backing support do not require safety glazing. Louvered windows or *jalousies* must meet certain dimension and material requirements but they also do not require safety glazing.

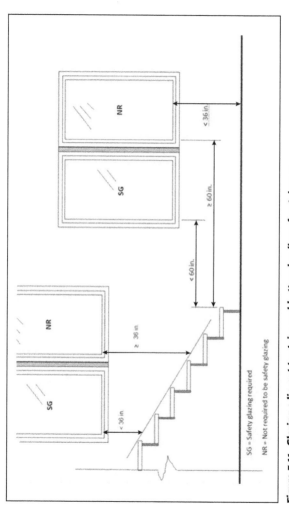

SG = Safety glazing required

NR = Not required to be safety glazing

Figure 3.11 Glazing adjacent to stairs and bottom landing of a stairway

Skylights

A skylight must have approved glazing materials, and in some cases a special retaining screen, depending on its height above the floor and the size and type of glazing material. Laminated glass larger than 16 sq. ft. or more than 12 ft. above the floor requires an interlayer thickness of 0.030 in. Fully tempered glass without a special retaining screen is limited to either of the following conditions:

- Glass area no more than 16 sq. ft., with the highest point of the glass no more than 12 ft. above a walking surface, nominal glass thickness less than or equal to 3/16 in., and any outboard panes fully tempered, laminated, or wired glass
  or
- Glass area greater than 16 sq. ft., sloped no more than 30 degrees, with highest point of glass no more than 10 ft. above a walking surface

Protecting Wood from Decay

Locations subject to decay require naturally durable wood or wood treated with preservatives. The heartwoods of redwood and cedar are considered naturally durable and decay-resistant. Preservative treatment must comply with the American Wood

Protection Association standards and be listed for the use. Lumber and plywood that is preservative treated must bear an approved quality mark. The following locations require protection from decay:

- Wood joists less than 18 in. or wood girders less than 12 in. above exposed ground in crawl spaces
- Wood sills and framing on concrete or masonry foundation walls less than 8 in. above exposed ground
- Plates on a concrete slab that is in direct contact with the ground without a moisture barrier
- Wood girders entering exterior masonry or concrete walls with clearances of less than ½ in. between the wood and the masonry or concrete on tops, sides, and ends of the girder
- Exterior wood siding, sheathing, and wall framing with less than 6 in. clearance from the ground or less than 2 in. above concrete slabs exposed to the weather
- Wood furring strips attached directly to the interior of exterior masonry or concrete walls below grade, except where an approved vapor retarder is applied between the wall and the furring strips

Ground Contact

Pressure-preservative-treated wood suitable for ground contact is required for structural supports in the following three locations (naturally durable wood is not permitted):

- In contact with the ground
- Embedded in concrete in contact with the ground
- Embedded in concrete exposed to the weather

Approved naturally durable or pressure-preservative-treated wood may be required for other structural members exposed to the weather, depending on geographic area.

Fasteners

Fasteners for pressure-preservative-treated wood shall be of hot-dipped, zinc-coated galvanized steel, stainless steel, silicon bronze, or copper. Such protection is not required for steel bolts of ½ in. diameter or larger, such as foundation anchor bolts.

Protection Against Termites

In areas subject to termite damage, wood structural elements must be protected by one or more of the following methods:

- Chemical termite treatment
- Termite-baiting system
- Pressure-preservative-treated wood
- Naturally durable termite-resistant wood

- Physical barriers (shields placed on top of exterior foundation walls are permitted only in combination with another method of protection)

Foam Plastic Protection

In likely termite infestation areas, the code prohibits the installation of foam plastics on foundations below grade. At least 6 in. clearance between foam plastics installed above grade and exposed earth is required. An approved method of protecting the foam plastic and structure from subterranean termite damage may be used as an alternative. These protection measures do not apply to the interior side of basement walls.

4

Foundations

The IRC contains prescriptive foundation designs to safely support building loads and transmit those loads to the soil. Builders should pay particular attention to the soil's characteristics and bearing capacity. The code bases footing size on soil-bearing capacity and the average gravity loads (dead, live, and snow) to be supported. As the load-bearing pressure of the soil decreases, footing size increases to distribute the load to a greater area. Fill soils that support footings and foundations must be designed, installed, and tested according to accepted engineering practices.

Soils

When local records indicate that the building site is likely to have expansive, compressible, shifting soil, or other questionable soil characteristics that might damage the structure, the building official may require a test to determine the soil's characteristics. When a complete geotechnical evaluation is not required and not performed, the code assumes the load-bearing values in table 4.1. Highly organic soils (laden with decayed material from plants), clays, silts, and peat are not included in the table. They require geotechnical analysis and, along with other expansive soils, require an engineered foundation design in accordance with the IBC.

Concrete

The compressive strength of concrete is expressed in pounds per square inch (psi) after 28 days curing time. The code requires concrete to have a minimum compressive strength of 2,500 psi for most applications (table 4.2). Stronger concrete, often including entrained air, is specified in geographic areas subject to moderate or severe weathering potential when the concrete is exposed to the weather or is for a garage floor slab.

Footings

Unless an approved alternative method is used, the code typically requires continuous concrete footings to support exterior walls. The number of stories

Table 4.1 Presumptive load-bearing values and properties of soils

Unified soil classification system symbol	Soil description	Load bearing pressure (psf)	Drainage	Frost heave potential	Volume change potential expansion
GW	Well-graded gravels, gravel-sand mixtures, little or no fines	3,000	Good	Low	Low
GP	Poorly graded gravels or gravel-sand mixtures, little or no fines	3,000	Good	Low	Low
SW	Well-graded sands, gravelly sands, little or no fines	2,000	Good	Low	Low
SP	Poorly graded sands or gravelly sands, little or no fines	2,000	Good	Low	Low
GM	Silty gravels, gravel-sand-silt mixtures	2,000	Good	Medium	Low
SM	Silty sand, sand-silt mixtures	2,000	Good	Medium	Low
GC	Clayey gravels, gravel-sand-clay mixtures	2,000	Medium	Medium	Low

(continued)

69

Table 4.1 Presumptive load-bearing values and properties of soils *(continued)*

Unified soil classification system symbol	Soil description	Load bearing pressure (psf)	Drainage	Frost heave potential	Volume change potential expansion
SC	Clayey sands, sand-clay mixture	2,000	Medium	Medium	Low
ML	Inorganic silts and very fine sands, rock flour, silty or clayey fine sands or clayey silts with slight plasticity	1,500	Medium	High	Low
CL	Inorganic clays of low to medium plasticity, gravelly clays, sandy clays, silty clays, lean clays	1,500	Medium	Medium	Medium to Low
CH	Inorganic clays of high plasticity, fat clays	1,500	Poor	Medium	High
MH	Inorganic silts, micaceous or diatomaceous fine sandy or silty soils, elastic silts	1,500	Poor	High	High

Table 4.2 Minimum specified compressive strength of concrete

Type or location of concrete construction	Minimum specified compressive strength at 28 days in psi		
	Weathering potential		
	Negligible	Moderate	Severe
Basement walls, foundations, and other concrete not exposed to the weather	2,500	2,500	2,500*
Basement slabs and interior slabs on grade, except garage floor slabs	2,500	2,500	2,500*
Basement walls, foundation walls, exterior walls, and other vertical concrete work exposed to the weather	2,500	3,000†	3,000†
Porches, carport slabs, and steps exposed to the weather, and garage floor slabs	2,500	3,000†	3,500†

* Concrete subject to freezing and thawing during construction shall be air-entrained.
† Concrete shall be air-entrained.

supported, the load-bearing value of the soil, the snow load, the type of construction and the type of foundation determine the minimum width and thickness of spread footings (table 4.3). The minimum footing thickness (T) is 6 in. but this thickness increases when the footing projection (P)—the distance from the edge of the footing to the edge of the foundation wall it is supporting—exceeds 6 in.

Table 4.3 Minimum width and thickness for concrete footings (in.)

Light-frame construction

Story and type of structure	Load-bearing value of soil (psf)					
	1500	2000	2500	3000	3500	4000
1 story–slab-on-grade	12 × 6	12 × 6	12 × 6	12 × 6	12 × 6	12 × 6
1 story–with crawl space	13 × 6	12 × 6	12 × 6	12 × 6	12 × 6	12 × 6
1 story–plus basement	19 × 6	14 × 6	12 × 6	12 × 6	12 × 6	12 × 6
2 story–slab-on-grade	12 × 6	12 × 6	12 × 6	12 × 6	12 × 6	12 × 6
2 story–with crawl space	17 × 6	13 × 6	12 × 6	12 × 6	12 × 6	12 × 6
2 story–plus basement	23 × 6	17 × 6	14 × 6	12 × 6	12 × 6	12 × 6
3 story–slab-on-grade	15 × 6	12 × 6	12 × 6	12 × 6	12 × 6	12 × 6
3 story–with crawl space	20 × 6	15 × 6	12 × 6	12 × 6	12 × 6	12 × 6
3 story–plus basement	26 × 8	20 × 6	16 × 6	13 × 6	12 × 6	12 × 6

Light-frame construction with brick veneer

Story and type of structure	Load-bearing value of soil (psf)					
	1500	2000	2500	3000	3500	4000
1 story–slab-on-grade	12 × 6	12 × 6	12 × 6	12 × 6	12 × 6	12 × 6
1 story–with crawl space	16 × 6	12 × 6	12 × 6	12 × 6	12 × 6	12 × 6
1 story–plus basement	22 × 6	16 × 6	13 × 6	12 × 6	12 × 6	12 × 6
2 story–slab-on-grade	16 × 6	12 × 6	12 × 6	12 × 6	12 × 6	12 × 6
2 story–with crawl space	22 × 6	16 × 6	13 × 6	12 × 6	12 × 6	12 × 6
2 story–plus basement	27 × 9	21 × 6	16 × 6	14 × 6	12 × 6	12 × 6
3 story–slab-on-grade	21 × 6	16 × 6	13 × 6	12 × 6	12 × 6	12 × 6
3 story–with crawl space	27 × 8	20 × 6	16 × 6	13 × 6	12 × 6	12 × 6
3 story–plus basement	33 × 11	24 × 7	20 × 6	16 × 6	14 × 6	12 × 6

Note: Values are based on a snow load of 30 psf.

(P cannot exceed T). The soil type and total tributary load determine pier and column footing sizes (total load/psf soil bearing pressure = minimum sq. ft. of footing).

Footing Placement

To protect a building's structural integrity, footings must meet the following conditions (fig. 4.1):

- Bear on undisturbed natural soils or engineered fill
- Be located at least 12 in. below undisturbed ground for exterior footings
- Not bear on frozen soil (unless frozen condition is permanent)
- Be below frost depth (unless there is a frost-protected shallow foundation system)
- Have a slope of 10% or less on the bottom of the footing

Note: The minimum width W is based on the soil bearing values in table 4.1 and the conditions in table 4.3. The footing projection P must be at least 2 in. and shall not exceed the footing thickness.

Frost protection is not required for the following structures:

- Light-framed accessory buildings of 600 sq. ft. or less, with an eave height of not more than 10 ft.

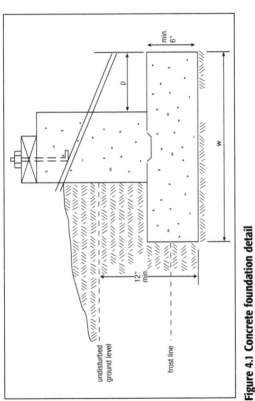

Figure 4.1 Concrete foundation detail

Source: 2006 International Residential Code Study Companion. Washington, DC: International Code Council, 2006, p.113.

Note: The minimum width W is based on the soil bearing values in table 4.1 and the conditions in table 4.3. The footing projection P must be at least 2 in. and shall not exceed the footing thickness.

- Other accessory buildings of 400 sq. ft. or less with an eave height of not more than 10 ft.
- Decks not supported by a dwelling

Concrete and Masonry Foundation Walls

The IRC provides prescriptive designs for cast-in-place concrete, precast concrete, and concrete masonry unit foundations supporting light-frame construction.

Cast-in-place Concrete Foundations

In general, the provisions for wall thickness, height, and reinforcement are the same for removable form and stay-in-place form concrete walls. Concrete basement walls must be supported laterally at the top and bottom before backfilling. The amount of vertical reinforcement is based on the following:

- Wall thickness
- Wall height
- Unbalanced backfill height
- Soil type

The minimum amount of horizontal reinforcing steel for cast-in-place concrete basement walls is one No. 4 bar at the top of the wall and the following:

- One No. 4 bar near mid-height of basement walls not more than 8 ft. high

  or

- One No. 4 bar near the ⅓ points of basement walls more than 8 ft. high

The code places additional limits on concrete walls located in Seismic Design Categories D_0, D_1, and D_2.

Precast Concrete Foundation Walls

Precast foundations are engineered systems that are manufactured in a controlled environment in various profiles including stud-and-cavity, and solid, composite, and hollow-core panels. They require concrete with a higher compressive strength. Precast foundations must be installed according to the manufacturer's instructions. The code prescribes the following footings:

- Concrete or solid masonry with continuous non-shrink grout

  or

- Compacted crushed stone (permitted in Seismic Design Categories A, B, and C)

The code provides tables for sizing crushed stone footings.

Masonry Foundation Walls

The code limits masonry foundation walls without reinforcing to 9 ft. or less, depending on the soil type, masonry thickness, and grouting of hollow units. The code generally does not permit masonry without reinforcing in Seismic Design Categories D_0, D_1, and D_2. Masonry basement walls must be laterally supported at the top and bottom before backfilling. The minimum amount of vertical reinforcement for masonry basement walls is based on

- Nominal masonry thickness of 8, 10, or 12 in.
- Wall height
- Unbalanced backfill height
- Soil type

The code places additional requirements on masonry walls located in Seismic Design Categories D_0, D_1, and D_2.

Surface and Foundation Drainage

Moving water away from the foundation is important to prevent moisture damage to a structure.

Grading for Surface Drainage

The IRC sets minimum grade requirements to adequately drain surface water. Concrete and masonry foundation walls must extend above the finished grade at least as high as follows:

- 6 in. when supporting wood frame walls
- 4 in. where the construction is masonry veneer over wood frame walls

Lots must be graded to drain surface water away from foundation walls to a storm sewer or other approved point of collection. Generally, grade must fall a minimum of 6 in. within the first 10 ft. from the building. When distances to property lines or other lot characteristics prohibit the prescribed slope, the code permits construction of swales or drains and does not specify a minimum slope.

Foundation Drainage

Drains are required around foundations that enclose habitable or usable spaces located below grade (fig. 4.2). Approved drainage systems must

- Be installed at or below the area to be protected
- Discharge by gravity or mechanical means to an approved location

Drainage tiles or perforated pipe must be

- Placed on at least 2 in. of washed gravel
- Covered with at least 6 in. of washed gravel

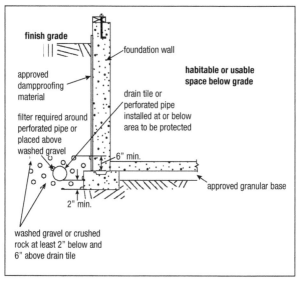

Figure 4.2 Foundation drain

Note: Perforated pipe requires a filter membrane around the pipe or over the gravel cover.

Gravel or crushed stone drains must
- Extend 12 in. beyond the edge of the footing
- Extend 6 in. above the footing
- Be covered with an approved filter membrane

Note: No drainage system is required when the foundation is installed on well-drained ground or sand-gravel mixture soils.

Dampproofing and Waterproofing

Approved coatings or membranes must be applied to the exterior of basement walls to prevent moisture penetration.

- Concrete and masonry foundations that enclose interior spaces and floors below grade must be dampproofed with approved materials from the top of the footing to the finished grade.
- Areas with a high water table or other known severe soil-water conditions require waterproofing.

Under-Floor Space

The code stipulates certain moisture control and access requirements for enclosed under-floor spaces, commonly referred to as crawl spaces.

Crawl Space Ventilation

The IRC prescribes the location and screening requirements for ventilation openings through foundation walls or exterior walls as follows:

- Install one ventilation opening within 3 ft. of each corner.
- Cover openings with approved vent, mesh, or screen materials.

The total net area of ventilation openings is as follows:
- With vapor retarder on ground surface, at least 1 sq. ft. for each 1,500 sq. ft. of area
- Without vapor retarder, at least 1 sq. ft. for each 150 sq. ft. of area

Unvented Crawl Space
For an unvented crawl space, the code requires the following:
- A vapor retarder cover on top of exposed earth
- Vapor retarder sealed to the stem wall
- Continuous mechanical exhaust ventilation or conditioned air supply provided
- Insulation installed on exterior walls

Access to Crawl Space
- Access openings through the floor must be at least 18 × 24 in.
- Openings through a perimeter wall must be at least 16 × 24 in.

5

Floors

This chapter addresses load-bearing requirements and floor framing details for both wood and steel floor framing, cantilevers, the use of wood structural panel floor sheathing, details for ground-level concrete floors, fire protection of light-frame construction, draft stopping and deck attachment.

Wood Floor Framing

Load-bearing dimension lumber for joists, beams, and girders must be identified by a grade mark. The intended use of the floor space, grade and species of lumber, and dead load in accordance with the appropriate span table determine allowable joist spans. For sleeping areas, habitable attics, and attics accessed by fixed stairs, the minimum live load is 30 psf. Other floor joists require a live load of not less than 40 psf. Floors of conventional wood construction typically have a dead load of 10 psf. For heavier flooring materials, joists may need to support a dead load of 20 psf. Allowable deflection for floor joists is limited to the length divided by 360 (L/360), as discussed in chapter 1. The joist spacing and spans (tables 5.1 and 5.2) account for this deflection.

Cantilever Joists for Wood Floor Framing

The code permits using prescriptive tables to determine the maximum span for floor cantilever joists when the joists support

- A light-frame bearing wall and roof only
- An exterior balcony

Table 5.3 assumes No. 2 grade of Douglas fir-larch, hem-fir, or spruce-pine-fir, or No. 1 grade southern pine for repetitive (3 or more) members. Uplift at the back span and connections capable of resisting the uplift forces also must be considered. The code requires a full-depth rim joist at the cantilevered end of the joists and solid blocking at the cantilever support.

Table 5.4 assumes No. 2 grade of Douglas fir-larch, hem-fir, or spruce-pine-fir, or No. 1 grade southern pine for repetitive (3 or more) members. The ratio of back span to cantilever must be at least 2:1. Uplift at the back span and connections capable of resisting the uplift forces also must be considered. The code requires a full-depth rim joist at the cantilevered end of the joists and solid blocking at the cantilever support.

Girders

Tables 6.4 and 6.5 in chapter 6 provide allowable spans for girders and headers fabricated of dimension lumber and located in an exterior or interior bearing wall. These tables

- Provide the required number of jack studs to support each end
- Consider the overall width of the building and the elements being supported in determining spans.
- Include values for ground snow loads of 30, 50, and 70 psf

Wood Floor Framing Details

Floor framing, including joists, headers, beams, and girders, shall be positively fastened to transfer all loads to the foundation and ensure against uplift and lateral displacement.

Joists Under Bearing Partitions

The IRC provides specific details for joists supporting bearing partitions as follows:

Table 5.1 Floor joist spans for common lumber species, #2 grade

(residential sleeping areas, live load = 30 psf, L/360)

Joist spacing (in.)	Species	Dead load = 10 psf			
		2 × 6	2 × 8	2 × 10	2 × 12
		Maximum floor joist spans			
		(ft.-in.)	(ft.-in.)	(ft.-in.)	(ft.-in.)
12	Douglas fir-larch	11-10	15-7	19-10	23-4
	Hem-fir	11-0	14-6	18-6	22-6
	Southern pine	11-3	14-11	18-1	21-4
	Spruce-pine-fir	11-3	14-11	19-0	23-0
16	Douglas fir-larch	10-9	14-2	17-5	20-3
	Hem-fir	10-0	13-2	16-10	19-8
	Southern pine	10-3	13-3	15-8	18-6
	Spruce-pine-fir	10-3	13-6	17-2	19-11
19.2	Douglas fir-larch	10-1	13-0	15-11	18-6
	Hem-fir	9-5	12-5	15-6	17-1
	Southern pine	9-6	12-1	14-4	16-10
	Spruce-pine-fir	9-8	12-9	15-8	18-3
24	Douglas fir-larch	9-3	11-8	14-3	16-6
	Hem-fir	8-9	11-4	13-10	16-1
	Southern pine	8-6	10-10	12-10	15-1
	Spruce-pine-fir	8-11	11-6	14-1	16-3

Table 5.1 Floor joist spans for common lumber species, #2 grade (*continued*)

(residential sleeping areas, live load = 30 psf, L/360)

Joist spacing (in.)	Species	Dead load = 20 psf			
		2 × 6	2 × 8	2 × 10	2 × 12
		Maximum floor joist spans			
		(ft.-in.)	(ft.-in.)	(ft.-in.)	(ft.-in.)
12	Douglas fir-larch	11-8	14-9	18-0	20-11
	Hem-fir	11-0	14-4	17-6	20-4
	Southern pine	10-9	13-8	16-2	19-1
	Spruce-pine-fir	11-3	14-7	17-9	20-7
16	Douglas fir-larch	10-1	12-9	15-7	18-1
	Hem-fir	9-10	12-5	15-2	17-7
	Southern pine	9-4	11-10	14-0	16-6
	Spruce-pine-fir	9-11	12-7	15-5	17-10
19.2	Douglas fir-larch	9-3	11-8	14-3	16-6
	Hem-fir	8-11	11-4	13-10	16-1
	Southern pine	8-6	10-10	12-10	15-1
	Spruce-pine-fir	9-1	11-6	14-1	16-3
24	Douglas fir-larch	8-3	10-5	12-9	14-9
	Hem-fir	8-0	10-2	12-5	14-4
	Southern pine	7-7	9-8	11-5	13-6
	Spruce-pine-fir	8-1	10-3	12-7	14-7

Table 5.2 Floor joist spans for common lumber species, #2 grade

(residential living areas, live load = 40 psf, L/360)

Joist spacing (in.)	Species and grade	Dead load = 10 psf			
		2 × 6	2 × 8	2 × 10	2 × 12
		Maximum floor joist spans			
		(ft.-in.)	(ft.-in.)	(ft.-in.)	(ft.-in.)
12	Douglas fir-larch	10-9	14-2	18-0	20-11
	Hem-fir	10-0	13-2	16-10	20-4
	Southern pine	10-3	13-6	16-2	19-1
	Spruce-pine-fir	10-3	13-6	17-3	20-7
16	Douglas fir-larch	9-9	12-9	15-7	18-1
	Hem-fir	9-1	12-0	15-2	17-7
	Southern pine	9-4	11-10	14-0	16-6
	Spruce-pine-fir	9-4	12-3	15-5	17-10
19.2	Douglas fir-larch	9-2	11-8	14-3	16-6
	Hem-fir	8-7	11-3	13-10	16-1
	Southern pine	8-6	10-10	12-10	15-1
	Spruce-pine-fir	8-9	11-6	14-1	16-3
24	Douglas fir-larch	8-3	10-5	12-9	14-9
	Hem-fir	7-11	10-2	12-5	14-4
	Southern pine	7-7	9-8	11-5	13-6
	Spruce-pine-fir	8-1	10-3	12-7	14-7

Table 5.2 Floor joist spans for common lumber species, #2 grade (*continued*)

(residential living areas, live load = 40 psf, L/360)

Joist spacing (in.)	Species and grade	Dead load = 20 psf			
		2 × 6	2 × 8	2 × 10	2 × 12
		Maximum floor joist spans			
		(ft.-in.)	(ft.-in.)	(ft.-in.)	(ft.-in.)
12	Douglas fir-larch	10-8	13-6	16-5	19-1
	Hem-fir	10-0	13-1	16-0	18-6
	Southern pine	9-10	12-6	14-9	17-5
	Spruce-pine-fir	10-3	13-3	16-3	18-10
16	Douglas fir-larch	9-3	11-8	14-3	16-6
	Hem-fir	8-11	11-4	13-10	16-1
	Southern pine	8-6	10-10	12-10	15-1
	Spruce-pine-fir	9-1	11-6	14-1	16-3
19.2	Douglas fir-larch	8-5	10-8	13-0	15-1
	Hem-fir	8-2	10-4	12-8	14-8
	Southern pine	7-9	9-10	11-8	13-9
	Spruce-pine-fir	8-3	10-6	12-10	14-10
24	Douglas fir-larch	7-6	9-6	11-8	13-6
	Hem-fir	7-4	9-3	11-4	13-1
	Southern pine	7-0	8-10	10-5	12-4
	Spruce-pine-fir	7-5	9-5	11-6	13-4

Table 5.3 Cantilever spans for floor joists supporting light-frame exterior bearing wall and roof only (floor live load ≤ 40 psf, roof live load ≤ 20 psf)

Member & spacing (in.)	≤ 20 psf Roof width			30 psf Roof width			50 psf Roof width			70 psf Roof width		
	24 ft.	32 ft.	40 ft.	24 ft.	32 ft.	40 ft.	24 ft.	32 ft.	40 ft.	24 ft.	32 ft.	40 ft.
2 × 8 @ 12	20	15	–	18	–	–	–	–	–	–	–	–
2 × 10 @ 16	29	21	16	26	18	–	20	–	–	–	–	–
2 × 10 @ 12	36	26	20	34	22	16	26	–	–	19	–	–
2 × 12 @ 16	–	32	25	36	29	21	29	20	–	23	–	–
2 × 12 @ 12	–	42	31	–	37	27	36	27	17	31	19	–
2 × 12 @ 8	–	48	45	–	48	38	–	40	26	36	29	18

Note: Spans are based on No. 2 grade lumber of Douglas fir-larch, hem-fir, and spruce-pine-fir for repetitive (three or more) members. No.1 or better shall be used for southern pine.

Table 5.4 Cantilever spans for floor joists supporting exterior balcony

Member size	Spacing (in.)	Maximum cantilever span (in.)		
		Ground snow load		
		≤ 30 psf	50 psf	70 psf
2 × 8	12	42	39	34
2 × 8	16	36	34	29
2 × 10	12	61	57	49
2 × 10	16	53	49	42
2 × 10	24	43	40	34
2 × 12	16	72	67	57
2 × 12	24	58	54	47

Note: Spans are based on No. 2 Grade lumber of Douglas fir-larch, hem-fir, and spruce-pine-fir for repetitive (three or more) members. No.1 or better shall be used for southern pine.

■ Joists under parallel bearing partitions must be adequate to support the load.

■ Double joists that are separated for installation of piping or vents and under parallel bearing partitions require full-depth solid blocking at 4 ft. *O.C.*

■ Bearing partitions perpendicular to joists shall not be offset by more than the joist depth from supporting girders or walls.

Bearing at End of Each Joist, Beam, or Girder

Joists, beams, and girders must be adequately supported at each end as follows:

- At least 1½ in. on wood or metal
- At least 3 in. on masonry or concrete

Alternative End Bearing Supports

Where the ends of joists do not bear on supporting walls or beams, the IRC allows support by one of the following methods:

- 1 × 4 ribbon strip, joist nailed to the adjacent stud
- Approved joist hangers

Joists Framing from Opposite Sides Over a Bearing Support

Where the ends of joists meet at a bearing support such as an interior beam, the joists must be

- Lapped at least 3 in. and nailed together with at least three 10d face nails

  or
- Spliced by wood or metal plate with strength equal to a nailed lap

Supports for Joists Framing into the Side of a Wood Girder

The code includes two methods of joist support at the sides of girders as follows:

- Approved framing anchors
  or
- 2 × 2 ledger strips

Lateral End Support for Joists

Any of the following methods will meet the IRC requirements for lateral support at the ends of floor joists:

- Full-depth solid blocking not less than 2 in. nominal thickness
- End of joist attached to a full-depth header, band, or rim joist
- End of joist attached to an adjoining stud
- Other approved lateral support to prevent rotation

Header and Trimmer Joist Requirements

For framing openings in the floor system, the code prescribes minimum strength and support requirements for headers, trimmers and tail joists based on the length of the wood framing members.

- Headers longer than 4 ft. must not be less than double members.
- Trimmers supporting headers longer than 4 ft. must not be less than double members.
- Headers and tail joists require hangers or other methods of support at bearing ends.

Cutting, Notching, and Drilling

Notches in lumber joists, rafters, and beams are limited as follows (fig. 5.1):

- Depth of no more than ⅙ the depth of the member (no more than ¼ at ends)
- Length of no more than ⅓ the depth of the member

Note: *Notches must not be located in the middle ⅓ of the span.*

Holes in joists, rafters, and beams are limited as follows:

- The diameter cannot be more than ⅓ the depth of the member.
- Holes must be located at least 2 in. from the top or bottom of the member.

Note: *Holes must be positioned at least 2 in. away from any other hole or notch.*

Cuts, notches, and holes in engineered wood products (trusses, *structural composite lumber,* structural glue-laminated members, or I-joists) are prohibited unless they are

- Permitted by the manufacturer's recommendations

  or

- Part of a specific designed alteration by a registered design professional

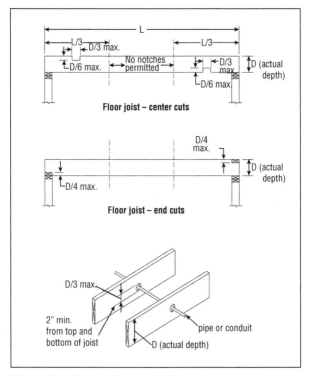

Figure 5.1 Cutting, notching, and drilling

Source: International Residential Code for One- and Two-Family Dwellings, 2006. Washington, DC: International Code Council, 2006, p. 105.

Wood Trusses

Trusses must be designed according to accepted engineering practice and referenced standards and braced in accordance with the individual truss design drawings. Truss design drawings must be

- Submitted to the building official and approved prior to installation
- Delivered to the jobsite with the trusses

Note: *Alterations to trusses are prohibited unless approved by a registered design professional.*

Draft Stopping

Draft stops are required to divide concealed floor spaces into approximately equal areas of no more than 1,000 sq. ft. These concealed spaces typically are created when a ceiling is applied to the bottom of open-web floor trusses or a ceiling is suspended below the floor joists.

Approved draft-stopping materials include the following:

- ½ in. gypsum board
- ⅜ in. wood structural panels
- ⅜ in. particleboard
- Other approved, adequately supported materials

Steel Floor Framing

The code contains prescriptive requirements for cold-formed steel floor framing applicable to buildings not greater than 40 × 60 ft. and not more than 3 stories high. Use of these code provisions is further limited to sites with an ultimate design wind speed of less than 139 mph in Exposure Category B or C and a maximum ground snow load of 70 psf.

Wood Structural Panel Floor Sheathing

Wood structural panels used for subfloor or combined subfloor/underlayment are typically plywood or oriented strand board (OSB). All panels must bear a grade mark issued by an approved agency. Maximum span ratings for application as roof and floor sheathing are part of this grade mark. For example, a grade mark of 24/16 would indicate a maximum roof rafter spacing of 24 in. and a maximum floor joist spacing of 16 in.

Combination subfloor/underlayment may have a single number indicating the span rating for floors only. The long dimension of the panel is applied perpendicular to the supports and the span rating is based on continuous panels over two or more spans. Requirements for edge blocking for other than tongue-and-groove structural panels depend on the grade and thickness of the panel and spacing of the supports.

Concrete Floors (on Ground)

Reinforcement must be supported to remain in place from the center to the upper ⅓ of the slab. Slabs must be at least 3.5 in. thick.

■ Vegetation, top soil, and foreign material must be removed.

■ Slabs below grade require 4-in.-thick sand or gravel base course.
   Note: Sand or gravel base course is not required for well-drained, sand-gravel soils.

■ A 6 mil polyethylene sheet or approved vapor retarder must be placed under the slab.
   Note: Unheated accessory structures or exterior areas do not require the vapor retarder.

Decks

The IRC provides prescriptive methods for deck construction including attachment to the structure as well as other connection and support details.

Deck Attachment

The code prescribes one method of attaching a deck ledger to the band or rim joist of the building structure (figs. 5.2 and 5.3). This prescriptive method applies only to the materials specified here:

- 2 × 8 or larger deck ledger of No. 2 preservative-treated lumber or approved decay-resistant lumber
- 2 in. nominal lumber band joist or 1 × 9½ Douglas fir laminated veneer lumber (LVL) rim board
- Maximum ½ in. wall sheathing

    Note: Maximum 1 in. sheathing is permitted with closer bolt spacing.

For deck attachment using the prescriptive methods, the IRC requires connection to the structure according to table 5.5 and as follows:

- Fasteners must not be less than ½ in. diameter lag screws or bolts with washers.
- Hot-dipped galvanized or stainless steel fasteners and washers must be used.
- Lag screws must be full depth through the rim joist.
- Fasteners must be staggered along the length of the ledger.
- Fasteners must be 2 in. minimum from the top of the ledger.

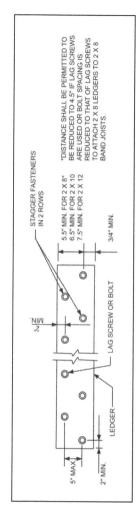

Figure 5.2 Placement of lag screws and bolts in ledgers

Source: International Residential Code for One-and Two-Family Dwelling. 2012. Washington, DC: International Code Council, 2011, p. 190.

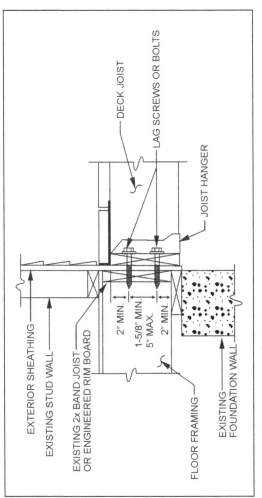

Figure 5.3 Deck ledger connection to structure

Source: International Residential Code for One-and Two-Family Dwelling, 2012. Washington, DC: International Code Council, 2011, p. 190.

Table 5.5 Deck ledger connection to band joist

(deck live load = 40 psf, deck dead load = 10 psf, snow load ≤ 40 psf)

Joist span	≤ 6'	6'1"–8'	8'1"–10'	10'1"–12'	12'1"–14'	14'1"–16'	16'1"–18'
Connection details			O.C. spacing of fasteners				
½" diameter lag screw with ½" maximum sheathing	30	23	18	15	13	11	10
½" diameter bolt with ½" maximum sheathing	36	36	34	29	24	21	19
½" diameter bolt with 1" maximum sheathing	36	36	29	24	21	18	16

- Fasteners must be 2 in. minimum from the bottom of the band joist.
- Fasteners must be 2 in. to 5 in. from the end of the ledger.
- Flashing must be placed over the ledger.

Other attachment methods must comply with accepted engineering practices.

Deck Joists and Beams

The IRC provides the maximum allowable spans for wood deck joists and beams using No. 2 grade of typical lumber species (tables 5.6 and 5.7).

The following information applies to wood deck joists and beams:

- Cantilevers limited to one-fourth of the joist or beam span
- Beam plies fastened with two rows of 10d nails at 16 in. O.C. along each edge
- Minimum 1 ½ in. bearing on wood or metal
- Minimum 3 in. bearing on concrete or masonry
- Support by approved joist hangers when framing into the side of a beam

Joists bearing on a beam require connection to resist lateral displacement. Two acceptable methods for attaching deck beams to deck posts are shown in figure 5.4. Other equivalent means may be used.

Table 5.6 Deck joist spans for common lumber species (ft.-in.)

Species	Size	Spacing of deck joists with no cantilever (in.)			Spacing of deck joists with cantilevers (in.)		
		12	16	24	12	16	24
Southern pine	2 × 6	9-11	9-0	7-7	6-8	6-8	6-8
	2 × 8	13-1	11-10	9-8	10-1	10-1	9-8
	2 × 10	16-2	14-0	11-5	14-6	14-0	11-5
	2 × 12	18-0	16-6	13-6	18-0	16-6	13-6
Douglas fir-larch, hem-fir spruce-pine-fir	2 × 6	9-6	8-8	7-2	6-3	6-3	6-3
	2 × 8	12-6	11-1	9-1	9-5	9-5	9-1
	2 × 10	15-8	13-7	11-1	13-7	13-7	11-1
	2 × 12	18-0	15-9	12-10	18-0	15-9	12-10
Redwood, western cedars, ponderosa pine, red pine	2 × 6	8-10	8-0	7-0	5-7	5-7	5-7
	2 × 8	11-8	10–7	8-8	8-6	8-6	8-6
	2 × 10	14-11	13–0	10-7	12-3	12-3	10-7
	2 × 12	17-5	15-1	12-4	16-5	15-1	12

Note: Spans are based on No. 2 grade lumber with wet service factor. Ground snow load, live load = 40 psf, dead load = 10 psf.

Table 5.7 Deck beam span lengths (ft.-in.)

Species	Size	Deck joist span (ft.)						
		≤ 6	≤ 8	≤ 10	≤ 12	≤ 14	≤ 16	≤ 18
Southern pine	2 – 2 × 6	6-11	5-11	5-4	4-10	4-6	4-3	4-0
	2 – 2 × 8	8-9	7-7	6-9	6-2	5-9	5-4	5-0
	2 – 2 × 10	10-4	9-0	8-0	7-4	6-9	6-4	6-0
	2 – 2 × 12	12-2	10-7	9-5	8-7	8-0	7-6	7-0
	3 – 2 × 6	8-2	7-5	6-8	6-1	5-8	5-3	5-0
	3 – 2 × 8	10-10	9-6	8-6	7-9	7-2	6-8	6-4
	3 – 2 × 10	13-0	11-3	10-0	9-2	8-6	7-11	7-6
	3 – 2 × 12	15-3	13-3	11-10	10-9	10-0	9-4	8-10

(continued)

Table 5.7 Deck beam span lengths (ft.-in.) (continued)

Species	Size	Deck joist span (ft.)								
		≤ 6	≤ 8	≤ 10	≤ 12	≤ 14	≤ 16	≤ 18
Douglas fir-larch, hem-fir, spruce-pine-fir, redwood, western cedars, ponderosa pine, red pine	3 × 6 or 2 – 2 × 6	5-5	4-8	4-2	3-10	3-6	3-1	2-9
	3 × 8 or 2 – 2 × 8	6-10	5-11	5-4	4-10	4-6	4-1	3-8
	3 × 10 or 2 – 2 × 10	8-4	7-3	6-6	5-11	5-6	5-1	4-8
	3 × 12 or 2 – 2 × 12	9-8	8-5	7-6	6-10	6-4	5-11	5-7
	4 × 6	6-5	5-6	4-11	4-6	4-2	3-11	3-8
	4 × 8	8-5	7-3	6-6	5-11	5-6	5-2	4-10
	4 × 10	9-11	8-7	7-8	7-0	6-6	6-1	5-8
	4 × 12	11-5	9-11	8-10	8-1	7-6	7-0	6-7
	3 – 2 × 6	7-4	6-8	6-0	5-6	5-1	4-9	4-6
	3 – 2 × 8	9-8	8-6	7-7	6-11	6-5	6-0	5-8
	3 – 2 × 10	12-0	10-5	9-4	8-6	7-10	7-4	6-11
	3 – 2 × 12	13-11	12-1	10-9	9-10	9-1	8-6	8-1

Note: Spans are based on No. 2 grade lumber with wet service factor. Ground snow load, live load = 40 psf, dead load = 10 psf. Beams supporting deck joists from one side only.

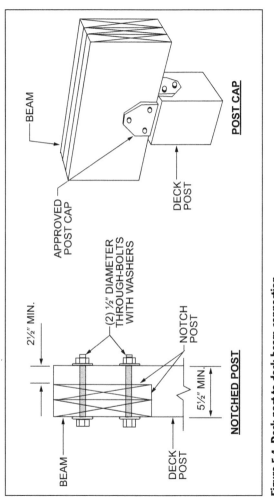

Figure 5.4 Deck post to deck beam connection

Source: International Residential Code for One- and Two-Family Dwellings, 2015. Washington, DC: International Code Council, 2014, p. 155.

Deck Posts

The code sets the maximum post height for supporting beams of single-level wood-framed decks (table 5.8).

Posts shall bear on footings and be restrained to prevent lateral displacement by one of the following:

- Manufactured connectors, or
- Minimum 12 in. embedment in soil or concrete

Table 5.8 Deck post height

Deck post size	Maximum height (ft.)
4 × 4	8
4 × 6	8
6 × 6	14

Note: Height is measured to the underside of the beam.

6

Wall Construction

Conventional light-frame wall construction, with its connections and bracing, must resist the code-prescribed vertical and lateral forces that act on a building, and transfer these loads to the foundation. In addition to setting limits on the size, length, and spacing of studs, the code addresses minimum connection requirements (tables 6.1 and 6.2). It presents a number of methods for wall bracing, which are critical to a building's structural integrity. This chapter also covers the installation of exterior doors and windows.

Wood Wall Framing

The prescriptive provisions of the IRC generally limit stud height in bearing walls to 10 ft. The required size and spacing of studs depend on the number of floors to be supported in addition to the roof (table 6.3).

Table 6.1 Fastening schedule

Item	Description of building elements	Number and type of fastener	Spacing and location
Roof			
1	Blocking between ceiling joists or rafters to top plate	4-8d box (2½" × 0.113"); or 3-8d common (2½" × 0.131"); or 3-10d box (3" × 0.128"); or 3-3" × 0.131" nails	Toe nail
2	Ceiling joists to top plate	4-8d box (2½" × 0.113"); or 3-8d common (2½" × 0.131"); or 3-10d box (3" × 0.128"); or 3-3" × 0.131" nails	Per joist, toe nail
3	Ceiling joist not attached to parallel rafter, laps over partitions, face nail	4-10d box (3" × 0.128"); or 3-16d common (3½" × 0.162"); or 4-3" × 0.131" nails	Face nail
4	Ceiling joist attached to parallel rafter (heel joint)	Per IRC Table R802.5.1(9)	Face nail
5	Collar tie to rafter, face nail or 1¼" × 20 gage ridge strap to rafter	4-10d box (3" × 0.128"); or 3-10d common (3" × 0.148"); or 4-3" × 0.131" nails	Face nail each rafter
6	Rafter or roof truss to plate	3-16d box nails (3½" × 0.135"); or 3-10d common nails (3" × 0.148"); or 4-10d box (3" × 0.128"); or 4-3" × 0.131" nails	2 toe nails on one side and 1 toe nail on opposite side of each rafter or truss

Table 6.1 Fastening schedule (*continued*)

Item	Description of building elements	Number and type of fastener	Spacing and location
Roof (*continued*)			
7	Roof rafters to ridge, valley or hip rafters or roof rafter to minimum 2" ridge beam	4-16d box (3½" × 0.135"); or 3-10d common (3½" × 0.148"); or 4-10d box (3" × 0.128"); or 4-3" × 0.131" nails	Toe nail
		3-16d box (3½" × 0.135") 2-16d common (3½" × 0.162"); or 3-10d box (3" × 0.128"); or 3-3" × 0.131" nails	End nail
Wall			
8	Stud to stud (not at braced wall panels)	16d common (3½" × 0.162")	24" O.C. face nail
		10d box (3" × 0.128"); or 3" × 0.131" nails	16" O.C. face nail
9	Stud to stud and abutting studs at intersecting wall corners (at braced wall panels)	16d box (3½" × 0.135"); or 3" × 0.131" nails	12" O.C. face nail
		16d common (3½" × 0.162")	16" O.C. face nail
10	Built-up header (2" to 2" header with ½" spacer)	16d common (3½" × 0.162")	16" O.C. along each edge face nail
		16d box (3½" × 0.135")	12" O.C. along each edge face nail

(*continued*)

Table 6.1 Fastening schedule (*continued*)

Item	Description of building elements	Number and type of fastener	Spacing and location
Wall (*continued*)			
11	Continuous header to stud	5-8d box (2½" × 0.113"); or 4-8d common (2½" × 0.131"); or 4-10d box (3" × 0.128")	Toe nail
12	Top plate to top plate	16d common (3½" × 0.162")	16" O.C. face nail
		10d box (3" × 0.128"); or 3" × 0.131" nails	12" O.C. face nail
13	Double top plate splice for SDCs A-D₂ with seismic braced wall line spacing < 25'	8-16d common (3½" × 0.162"); or 12-16d box (3½" × 0.135"); or 12-10d box (3" × 0.128"); or 12-3" × 0.131" nails	Face nail on each side of end joint (minimum 24" lap splice length each side of end joint)
	Double top plate splice SDCs D₀, D₁, or D₂; and braced wall line spacing ≥ 25'	12-16d (3½" × 0.135")	
14	Bottom plate to joist, rim joist, band joist or blocking (not at braced wall panels)	16d common (3½" × 0.162")	16" O.C. face nail
		16d box (3½" × 0.135"); or 3" × 0.131" nails	12" O.C. face nail
15	Bottom plate to joist, rim joist, band joist or blocking (at braced wall panel)	3-16d box (3½" × 0.135"); or 2-16d common (3½" × 0.162"); or 4-3" × 0.131" nails	16" O.C. face nail

Table 6.1 Fastening schedule (*continued*)

Item	Description of building elements	Number and type of fastener	Spacing and location
	Wall (*continued*)		
16	Top or bottom plate to stud	4-8d box (2½" × 0.113");or 3-16d box (3½" × 0.135"); or 4-8d common (2½" × 0.131"); or 4-10d box (3" × 0.128"); or 4-3" × 0.131" nails	Toe nail
17	Top or bottom plate to stud	3-16d box (3½" × 0.135"); or 2-16d common (3½" × 0.162"); or 3-10d box (3" × 0.128"); or 3-3" × 0.131" nails	End nail
18	Top plates, laps at corners and intersections	3-10d box (3" × 0.128"); or 2-16d common (3½" × 0.162"); or 3-3" × 0.131" nails	Face nail
19	1" brace to each stud and plate	3-8d box (2½" × 0.113"); or 2-8d common (2½" × 0.131"); or 2-10d box (3" × 0.128") 2 staples 1¾"	Face nail
	Floor		
20	Joist to sill, top plate or girder	4-8d box (2½" × 0.113"); or 3-8d common (2½" × 0.131"); or 3-10d box (3" × 0.128"); or 3-3" × 0.131" nails	Toe nail
21	Rim joist, band joist or blocking to sill or top plate (roof applications also)	8d box (2½" × 0.113")	4" O.C. toe nail
		8d common (2½" × 0.131"); or 10d box (3" × 0.128"); or 3" × 0.131" nails	6" O.C. toe nail

(continued)

Table 6.1 Fastening schedule (*continued*)

Item	Description of building elements	Number and type of fastener	Spacing and location
Floor (*continued*)			
22	Band or rim joist to joist	3-16d common (3½" × 0.162") or 4-10 box (3" × 0.128), or 4-3" × 0.131" nails, or 4-3" × 14 gage staples, 7⁄16" crown	End nail
23	Built-up girders and beams, 2" lumber layers	20d common (4" × 0.192"); or	Nail each layer as follows: 32" O.C. at top and bottom and staggered.
		10d (3" × 0.128"); or 3-3" × 0.131" nails	24" O.C. face nail at top and bottom staggered on opposite sides
		And: 2-20d common (4" × 0.192"); or 3-10d box (3" × 0.128"); or 3-3" × 0.131" nails	Face nail at ends and at each splice
24	Ledger strip supporting joists or rafters	4-16d box (3½" × 0.135"); or 3-16d common (3½" × 0.162"); or 4-10d box (3" × 0.128"); or 4-3" × 0.131" nails	At each joist or rafter, face nail
25	Joist to band joist or rim joist	4-10d (3" × 0.128")	End nail
26	Bridging to joist	2-10d (3" × 0.128")	Each end, toe nail

Table 6.1 Fastening schedule (*continued*)

Item	Description of building elements	Number and type of fastener	Spacing of fasteners Edges (in.)	Inter-mediate supports (in.)
Wood structural panels, subfloor, roof and interior wall sheathing to framing, and particleboard wall sheathing to framing				
27	⅜″–½″	6d common (2″ × 0.113″) nail (subfloor, wall) 8d common (2½″ × 0.131″) nail (roof)	6	12
28	¹⁹⁄₃₂″–1″	8d common nail (2½″ × 0.131″)	6	12
29	1⅛″–1¼″	10d common (3″ × 0.148″) nail or 8d (2½″ × 0.131″) deformed nail	6	12
Other wall sheathing				
30	½″ structural cellulosic fiber-board sheathing	1½″ galvanized roofing nail; ⁷⁄₁₆″ crown or 1″ crown staple 16 ga., 1¼″ long	3	6
31	²⁵⁄₃₂″ structural cellulosic fiber-board sheathing	1¾″ galvanized roofing nail; ⁷⁄₁₆″ crown or 1″ crown staple 16 ga., 1½″ long	3	6
32	½″ gypsum sheathing	1½″ galvanized roofing nail; staple galvanized, 1½″ long; 1¼″ screws, Type W or S	7	7
33	⅝″ gypsum sheathing	1¾″ galvanized roofing nail; staple galvanized, 1⅝″ long; 1⅝″ screws, Type W or S	7	7

(*continued*)

Table 6.1 Fastening schedule (*continued*)

Item	Description of building elements	Number and type of fastener	Edges (in.)	Inter-mediate supports (in.)
			Spacing of fasteners	
Wood structural panels, combination subfloor underlayment to framing				
34	¾" and less	6d deformed (2" × 0.120") nail or 8d common (2½" × 0.131") nail	6	12
35	⅞"–1"	8d common (2½" × 0.131") nail or 8d (2½" × 0.120") deformed nail	6	12
36	1⅛"–1¼"	10d common (3" × 0.148") nail or 8d (2½" × 0.120") deformed nail	6	12

Table 6.2 Requirements for wood structural panel wall sheathing used to resist wind pressures

Minimum nail		Minimum wood structural panel span rating	Minimum nominal panel thickness (in.)	Maximum wall stud spacing (in.)	Panel nail spacing		Ultimate design wind speed V$_{ult}$ (mph)		
Size	Penetration (in.)				Edges (in. O.C.)	Field (in. O.C.)	Wind exposure category		
							B	C	D
6d Common (2.0" × 0.113")	1.5	24/0	3/8	16	6	12	140	115	110
8d Common (2.5" × 0.131")	1.75	24/16	7/16	16	6	12	170	140	135
				24	6	12	140	115	110

Table 6.3 Size, height, and spacing of wood studs

Stud size (in.)	Bearing walls					Nonbearing walls	
	Laterally unsupported stud height (ft.)	Maximum spacing when supporting roof-ceiling assembly or a habitable attic assembly, only (in.)	Maximum spacing when supporting one floor, plus a roof-ceiling assembly or a habitable attic assembly (in.)	Maximum spacing when supporting two floors, plus a roof-ceiling assembly or a habitable attic assembly (in.)	Maximum spacing when supporting one floor only (in.)	Laterally unsupported stud height (ft.)	Maximum spacing (in.)
2 × 3	–	–	–	–	–	10	16
2 × 4	10	24	16	–	24	14	24
3 × 4	10	24	24	16	24	14	24
2 × 5	10	24	24	–	24	16	24
2 × 6	10	24	24	16	24	20	24

Note: A habitable attic assembly supported by 2" × 4" studs is limited to a roof span of 32 ft. Where the roof span exceeds 32 ft, the wall studs shall be increased to 2" x 6" or the studs shall be designed in accordance with accepted engineering practice.

118

Load-bearing dimension lumber for wall framing must meet the following requirements:

- Identified by a grade mark
- Studs No. 3, standard or stud-grade lumber
  Note: *Utility studs are permitted, provided they do not support floors.*

Double top plates are generally required for exterior walls and interior bearing walls. They must meet the following requirements:

- Overlap at corners and intersections with bearing partitions
- End joints offset at least 24 in.
  Note: *The code permits a single top plate when studs are in line with joists and rafters, and all plate joints are reinforced with additional connectors.*

In dwelling construction, interior and exterior wall studs typically are installed at 16 in. O.C. If the bearing studs are 24 in. O.C., and the joists, trusses, or rafters above them are more than 16 in. O.C., the framing must have one of the following:

- Bearing located within 5 in. of the stud below
- Two 2 × 6 or two 3 × 4 top plates
- A third top plate
- Solid blocking to reinforce double top plate

For interior nonbearing walls, the code permits 2 × 3 studs, 24 in. O.C. or 2 × 4 flat studs, 16 in. O.C.

Drilling and Notching

To preserve the structural integrity of the wall framing, the IRC limits the amount of notching and boring in studs and plates. Notches and bored holes in studs of exterior walls or bearing partitions must meet the following requirements (fig. 6.1):

- Maximum notch 25% of the stud depth
- Maximum hole diameter of 40% of the stud depth
- Edge of hole at least ⅝ in. from edge of stud

The code permits drilling of up to 60% diameter holes in studs of exterior walls or bearing partitions provided the studs are doubled, with no more than two successive doubled studs bored.

Notches and bored holes in studs of nonbearing partitions must meet the following requirements (fig. 6.2):

- Maximum notch of 40% of the stud depth
- Maximum hole diameter of 60% of the stud depth
- Edge of hole at least ⅝ in. from edge of stud

Unless wood structural panel sheathing covers the entire side of the wall, top plates of exterior or bearing walls that are drilled or notched more than 50% require reinforcement with a metal strap as follows (fig. 6.3):

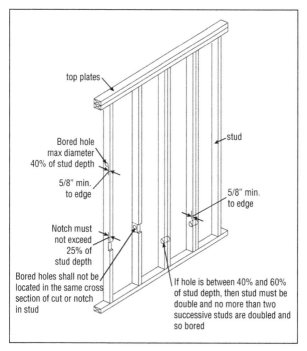

top plates

stud

Bored hole
max diameter
40% of stud depth

5/8" min.
to edge

5/8" min.
to edge

Notch must
not exceed
25% of
stud depth

Bored holes shall not be
located in the same cross
section of cut or notch
in stud

If hole is between 40% and 60%
of stud depth, then stud must be
double and no more than two
successive studs are doubled and
so bored

Figure 6.1 Notching and bored hole limitations for exterior walls and bearing walls

Source: International Residential Code for One- and Two-Family Dwellings, 2006. Washington, DC: International Code Council, 2006, p. 132

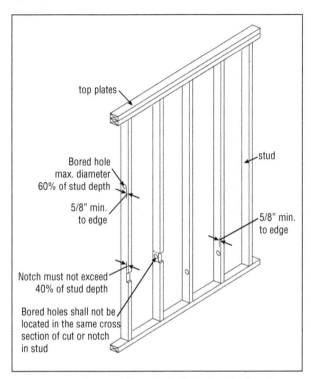

Figure 6.2 Notching and bored hole limitations for interior nonbearing walls

Source: International Residential Code for One- and Two-Family Dwellings, 2006. Washington, DC: International Code Council, 2006, p. 133

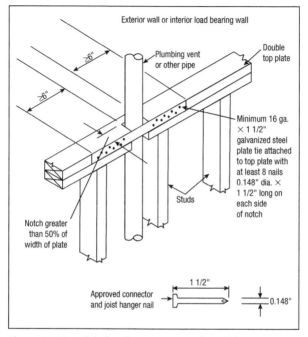

Exterior wall or interior load bearing wall

Plumbing vent
or other pipe

Double
top plate

≥6"

≥6"

Minimum 16 ga.
× 1 1/2"
galvanized steel
plate tie attached
to top plate with
at least 8 nails
0.148" dia. ×
1 1/2" long on
each side
of notch

Studs

Notch greater
than 50% of
width of plate

Approved connector
and joist hanger nail

1 1/2"

0.148"

Figure 6.3 Top plate framing to accommodate piping

- Galvanized metal tie 0.054 in. thick (16 gage) and 1½ in. wide
- Metal tie 6 in. past opening on each side
- Eight joist hanger nails (0.148 in. diameter × 1½ in.) at each side

Headers

Tables 6.4 and 6.5 provide allowable spans for headers fabricated of dimension lumber and located in an exterior or interior bearing wall. These tables

- Provide the required number of jack studs to support each end
- Consider the overall width of the building and the elements being supported in determining spans
- Include values for ground snow loads of 30, 50, and 70 psf

   Note: The code also prescribes the minimum number of full-height studs at each end of the header based on the maximum header span.

Fireblocking

To stop the spread of fire in concealed spaces of wood frame construction, the IRC requires fireblocking to cut off all concealed draft openings and

Table 6.4 Girder spans and header spans for exterior bearing walls

(maximum spans for Douglas fir-larch, hem-fir, southern pine, and spruce-pine-fir and required number of jack studs)

Girders and headers supporting	Quantity	Size	Ground snow load (psf)											
			30				50				70			
			Building width (ft.)											
			28		36		28		36		28		36	
			Span	Jack studs	Span	Jack studs	Span	Jack studs	Span	Jack studs	Span	Jack studs	Span	Jack studs
Roof and ceiling	2	2 × 8	5-11	2	5-4	2	5-2	2	4-7	2	4-7	2	4-1	2
	2	2 × 10	7-3	2	6-6	2	6-3	2	5-7	2	5-7	2	5-0	2
	2	2 × 12	8-5	2	7-6	2	7-3	2	6-6	2	6-6	2	5-10	3
	3	2 × 8	7-5	1	6-8	1	6-5	2	5-9	2	5-9	2	5-2	2
	3	2 × 10	9-1	2	8-2	2	7-10	2	7-0	2	7-0	2	6-4	2
	3	2 × 12	10-7	2	9-5	2	9-2	2	8-2	2	8-2	2	7-4	2

(continued)

Table 6.4 Girder spans and header spans for exterior bearing walls (continued)

Girders and headers supporting	Quantity	Size	Ground snow load (psf)											
			30				50				70			
			Building width (ft.)											
			28		36		28		36		28		36	
			Span	Jack studs	Span	Jack studs	Span	Jack studs	Span	Jack studs	Span	Jack studs	Span	Jack studs
Roof, ceiling, and one center-bearing floor	2	2 × 8	5-0	2	4-6	2	4-6	2	4-1	2	4-2	2	3-9	2
	2	2 × 10	6-2	2	5-6	2	5-6	2	5-0	2	5-1	2	4-7	3
	2	2 × 12	7-1	2	6-5	2	6-5	2	5-9	3	5-10	3	5-3	3
	3	2 × 8	6-3	2	5-8	2	5-8	2	5-1	2	5-2	2	4-8	2
	3	2 × 10	7-8	2	6-11	2	6-11	2	6-3	2	6-4	2	5-8	2
	3	2 × 12	8-11	2	8-0	2	8-0	2	7-3	2	7-4	2	6-7	2
Roof, ceiling, and one clear span floor	2	2 × 8	4-4	2	3-10	2	4-2	2	3-9	2	3-11	2	3-6	2
	2	2 × 10	5-3	2	4-8	2	5-1	2	4-7	3	4-9	2	4-3	3
	2	2 × 12	6-1	3	5-5	3	5-11	3	5-4	3	5-6	3	5-0	3

Category		Size												
Roof, ceiling, and one clear span floor (continued)	3	2 × 8	5-5	2	4-10	2	5-3	2	4-8	2	4-11	2	4-5	2
	3	2 × 10	6-7	2	5-11	2	6-5	2	5-9	2	6-0	2	5-4	2
	3	2 × 12	7-8	2	6-10	2	7-5	2	6-8	2	6-11	2	6-3	2
	2	2 × 8	4-2	2	3-9	2	4-0	2	3-8	2	3-9	2	3-5	2
	2	2 × 10	5-1	2	4-7	3	4-11	2	4-5	3	4-7	3	4-2	3
Roof, ceiling, and two center-bearing floors	2	2 × 12	5-10	3	5-3	3	5-9	3	5-2	3	5-4	3	4-10	3
	3	2 × 8	5-2	2	4-8	2	5-1	2	4-7	2	4-9	2	4-3	2
	3	2 × 10	6-4	2	5-8	2	6-2	2	5-7	2	5-9	2	5-3	2
	3	2 × 12	7-4	2	6-7	2	7-2	2	6-5	2	6-9	2	6-1	3
	2	2 × 8	3-4	2	3-0	3	3-4	3	2-11	3	3-3	3	2-11	3
Roof, ceiling, and two clear-span floors	2	2 × 10	4-1	3	3-8	3	4-0	3	3-7	3	4-0	3	3-6	3
	2	2 × 12	4-9	3	4-3	3	4-8	3	4-2	3	4-7	3	4-1	4
	3	2 × 8	4-2	2	3-9	2	4-1	2	3-8	2	4-1	2	3-8	2
	3	2 × 10	5-1	2	4-7	3	5-0	3	4-6	3	4-11	2	4-5	2
	3	2 × 12	5-11	3	5-4	3	5-10	3	5-3	3	5-9	3	5-2	3

Note: Spans are based on No. 2 Grade lumber of Douglas fir-larch, hem-fir, and spruce-pine-fir. No.1 or better shall be used for southern pine.

Table 6.5 Girder spans and header spans for interior bearing walls

(maximum spans for Douglas fir-larch, hem-fir, southern pine, and spruce-pine-fir and required number of jack studs)

Headers and girders supporting	Quantity	Size	Building width (ft.)					
			20		28		36	
			Span	Jack studs	Span	Jack studs	Span	Jack studs
	2	2 × 4	3–1	1	2–8	1	2–5	1
	2	2 × 6	4–6	1	3–11	1	3–6	1
	2	2 × 8	5–9	1	5–0	2	4–5	2
	2	2 × 10	7–0	2	6–1	2	5–5	2
	2	2 × 12	8–1	2	7–0	2	6–3	2
One floor only	3	2 × 8	7–2	1	6–3	1	5–7	2
	3	2 × 10	8–9	1	7–7	2	6–9	2
	3	2 × 12	10–2	2	8–10	2	7–10	2
	4	2 × 8	9–0	1	7–8	1	6–9	1
	4	2 × 10	10–1	1	8–9	1	7–10	2
	4	2 × 12	11–9	1	10–2	2	9–1	2

		2-2	1	1-10	1	1-7	1
2	2 × 4	3-2	2	2-9	2	2-5	2
2	2 × 6	4-1	2	3-6	2	3-2	2
2	2 × 8	4-11	2	4-3	2	3-10	3
2	2 × 10	5-9	2	5-0	3	4-5	3
2	2 × 12	5-1	2	4-5	2	3-11	2
Two floors	3	2 × 8	6-2	2	5-4	2	4-10	2
3	2 × 10	7-2	2	6-3	2	5-7	3
3	2 × 12	6-1	1	5-3	2	4-8	2
4	2 × 8	7-2	2	6-2	2	5-6	2
4	2 × 10	8-4	2	7-2	2	6-5	2
4	2 × 12						

Note: Spans are based on No. 2 Grade lumber of Douglas fir-larch, hem-fir, and spruce-pine-fir. No.1 or better shall be used for southern pine.

to form an effective fire barrier between stories and between a top story and the roof space (attic). The following concealed locations require fireblocking:

- Walls at the ceiling and floor levels (by using top and bottom plates in platform framing)
- Walls horizontally at 10 ft. intervals or less (provided by studs)
- Interconnections between vertical and horizontal spaces, such as those at soffits, drop ceilings, and cove ceilings
- Concealed spaces between stair stringers at the top and bottom of the run
- Openings around vents, pipes, ducts, cables, and wires at ceiling and floor level
- Cornices of two-family dwellings at the separation line

Fireblocking Materials

The code requires an approved material, such as caulking, around vents, pipes, ducts, cables, and wires to resist passage of flame and products of combustion. In other locations, approved fireblocking materials include the following:

- 2 in. nominal lumber
- 2 thicknesses of 1 in. nominal lumber
- $\frac{23}{32}$ in. wood structural panel with joints backed by $\frac{23}{32}$ in. wood structural panel

- ¾ in. particleboard with joints backed by ¾ in. particleboard
- ½ in. gypsum board
- ¼ in. cement-based millboard
- Batts or blankets of mineral wool or glass fiber securely retained in place
- Cellulose insulation installed as tested for the specific application

Foundation Cripple Walls

A cripple wall is a framed wall extending from the top of the foundation to the underside of the floor framing of the first-story-above-*grade plane*. This wall must meet the following criteria:

- The studs in the cripple wall cannot be smaller than the studs that are above it.
- A stud height of less than 14 in. requires wood structural panel sheathing unless the cripple wall is constructed of solid blocking.
- Cripple walls shall be supported on continuous foundations.

Wall Bracing

Wall bracing is necessary to provide resistance to *racking* from lateral loads, primarily wind and seismic forces. The IRC includes various prescriptive methods of isolated panel and diagonal wall bracing, referred to as intermittent bracing methods. In addition to intermittent *braced wall panels* along a *braced wall line,* another, more common, method for compliance with the bracing requirements is continuous sheathing. This method applies wood structural panel or structural fiberboard sheathing to all surfaces along the braced wall line. The code includes several alternatives to both intermittent and continuous methods for narrow wall bracing, typically used at corners or adjacent to overhead door openings.

Braced Wall Lines

A braced wall line consists of a series of braced wall panels resisting lateral forces in the plane of the wall. For buildings sited in SDC D_0, D_1, or D_2, and townhomes in SDC C, the maximum spacing between braced wall lines typically is 25 ft. Under some circumstances, the code permits an increase to 35 ft. spacing. All buildings located in SDC A or B, and one- and two-family dwellings in SDC C are exempt from the seismic provisions and bracing is based on the design wind speed. In these locations, the spacing between braced wall lines corresponds to the amount of required bracing: greater spacing requires more bracing.

Intermittent Braced Wall Panel Construction Methods

The IRC recognizes 9 methods for constructing intermittent braced wall panels and 3 alternatives. The most common bracing methods are shown in table 6.6. The minimum length of a braced wall panel is typically 4 ft., but the alternatives permit a reduction in this length by increasing panel strength through specific material, connection, and anchorage details. In some cases, the code also permits partial credit for braced wall panels that are less than 4 ft. in actual length in SDC A, B, and C.

A braced wall line must contain the prescribed total length of bracing in feet (table 6.7) and meet the maximum spacing requirements. The amount and location of bracing is determined by various factors. These include:

- Number of stories
- SDC
- Design wind speed
- Wind exposure category
- Bracing method

Intermittent Braced Wall Panel

Intermittent braced wall panels within a braced wall line must

- Be spaced no more than 20 ft. between panels

Table 6.6 Bracing methods

Method	Material	Minimum thickness	Illustration	Connection criteria
LIB	Let-in-bracing	1 × 4 wood or approved metal straps at 45° to 60° angles		Wood: 2 8d common nails or 3 8d (2½″ long × 0.113″ dia.) nails per stud and top and bottom plates
WSP	Wood structural panel	³⁄₈″		For exterior sheathing see table 6.2 For interior sheathing see table 6.1
SFB	Structural fiberboard sheathing	½″ or ²⁵⁄₃₂″ for maximum 16″ stud spacing		1½″ long × 0.12″ dia. (for ½″ thick sheathing) 1¾″ long × 0.12″ dia. (for ²⁵⁄₃₂″ thick sheathing) galvanized roofing nails or 8d common (2½″ long × 0.131″ dia.) nails; 3″ edges; 6″ field

ABW	Alternate braced wall	3⁄8″		See Section R602.10.6.1.
PFH	Portal frame with hold-downs	3⁄8″		See figure 6.4.
PFG	Portal frame at garage	7⁄16″		See figure 6.5.
CS-WSP	Continuously sheathed wood structural panel	3⁄8″		For exterior sheathing see table 6.2. For interior sheathing see table 6.1.

(continued)

Table 6.6 Bracing methods *(continued)*

Method	Material	Minimum thickness	Illustration	Connection criteria
CS-G	Continuously sheathed wood panel structural adjacent to garage openings	3⁄8″		For exterior sheathing see table 6.2. For interior sheathing see table 6.1.
CS-PF	Continuously sheathed portal frame	7⁄16″		See figure 6.6.
CS-SFB	Continuously sheathed structural fiberboard	1⁄2″ or 25⁄32″ for maximum 16″ stud spacing		1½″ long × 0.12″ dia. (for ½″ thick sheathing) 1¾″ long × 0.12″ dia. (for 25⁄32″ thick sheathing) galvanized roofing nails or 8d common (2½″ long × 0.131″ dia.) nails; 3″ edges, 6″ field

Table 6.7 Bracing requirements based on wind speed

Story Location	Braced wall line spacing (ft.)	Minimum total length (ft.) of braced wall panels required along each braced wall line			
		Method LIB	Method GB	Methods DWB, WSP, SFB, PCP, HPS, CS-SFB	Methods CS-WSP, CS-G, CS-PF
	10	3.5	3.5	2	2
	20	6.5	6.5	3.5	3.5
	30	9.5	9.5	5.5	4.5
	40	12.5	12.5	7	6
	50	15	15	9	7.5
	60	18	18	10.5	9
	10	7	7	4	3.5
	20	12.5	12.5	7.5	6.5
	30	18	18	10.5	9
	40	23.5	23.5	13.5	11.5
	50	29	29	16.5	14
	60	34.5	34.5	20	17

(continued)

Table 6.7 Bracing requirements based on wind speed *(continued)*

Story Location	Braced wall line spacing (ft.)	Minimum total length (ft.) of braced wall panels required along each braced wall line			
		Method LIB	Method GB	Methods DWB, WSP, SFB, PBS, PCP, HPS, CS-SFB	Methods CS-WSP, CS-G, CS-PF
◁▢▢▨ ◁▢▢ ◁▢	10	NP	10	6	5
	20	NP	18.5	11	9
	30	NP	27	15.5	13
	40	NP	35	20	17
	50	NP	43	24.5	21
	60	NP	51	29	25

Bracing amounts are based on the following:
- Ultimate Design Wind Speed 115 mph
- Exposure Category B
- 30 ft mean roof height
- 10 ft eave to ridge height
- 10 ft wall height
- 2 braced wall lines

138

- Not be offset out of plane of the braced wall line by more than 4 ft. (8 ft. total out-to-out offset)
- Start within 10 ft. of either end of braced wall line

  Note: *If continuous sheathing is not used, braced wall panels must be located at each end of the braced wall line in seismic design categories D_0, D_1, and D_2 with some exceptions.*

Continuous Wood Structural Panel Sheathing

Another way (and typically the most common method) to comply with the wall bracing provisions is to apply wood structural panels to all areas of one side of a braced wall line, including above and below windows. This continuous sheathing method increases the rigidity of the lateral resistance system and allows reduced lengths for full-height braced wall panels. *(See table 6.7 for total required length of bracing in a braced wall line using continuous sheathing and based on an ultimate design wind speed of 115 mph.)* The code permits mixing continuous and intermittent bracing methods with some restrictions based on seismic design category and wind speed. The minimum length of a braced wall panel is based on the height of the adjacent window or door opening (table 6.8).

When using wood structural panels, continuous sheathing is known as Method CS-WSP. The code

Table 6.8 Minimum length of braced wall panels for continuous sheathing (methods CS-WSP, CS-SFB)

Adjacent clear opening height (in.)	Wall height (ft.)				
	8	9	10	11	12
64	24	27	30	33	36
68	26	27	30	33	36
72	27	27	30	33	36
76	30	29	30	33	36
80	32	30	30	33	36
84	35	32	32	33	36
88	38	35	33	33	36
92	43	37	35	35	36
96	48	41	38	36	36
100	–	44	40	38	38
104	–	49	43	40	39
108	–	54	46	43	41
112	–	–	50	45	43
116	–	–	55	48	45
120	–	–	60	52	48
124	–	–	–	56	51
128	–	–	–	61	54
132	–	–	–	66	58
136	–	–	–	–	62
140	–	–	–	–	66
144	–	–	–	–	72

also permits continuous structural fiberboard sheathing (Method CS-FSB). The continuous sheathing methods including alternatives are shown in table 6.6.

Alternate and Portal Frame Braced Wall Panels

The code provides alternative designs, including 3 portal frame methods, for reducing braced wall panel lengths by increasing the strength of the panel through specific material, connection, and anchorage details. The minimum length of the panel for the portal frame methods is based on wall height (table 6.9). Table 6.9 also determines the contributing

Table 6.9 Minimum length of braced wall panels for portal frame methods

| Method | | Minimum length (in.) | | | | | Contributing length (in.) |
| | | Wall height | | | | | |
		8′	9′	10′	11′	12′	
PFH	Supporting roof only	16	16	16	18	20	48
	Supporting one story and roof	24	24	24	27	29	48
PFG		24	27	30	33	36	1.5 × actual
CS-PF	SDC A, B and C	16	18	20	22	24	1.5 × actual
	SDC D_0, D_1, and D_2	16	18	20	22	24	actual

length of each alternative panel to the required total length of bracing in the wall line.

Method PFH: Portal Frame with Hold-Downs

Method PFH requires hold-down devices for anchorage to the foundation and may be used in all seismic design categories within the scope of the IRC. This is the only narrow wall method that receives full credit as equivalent to a 48 in. braced wall panel. Construction must comply with the details of figure 6.4 as follows:

- Panel length is determined from table 6.9, but must be at least
  - 16 in. for 1-story buildings
  - 24 in. for the first story of 2-story buildings
- Wall height not more than 12 ft.
- Top of header height not more than 10 ft.
- Minimum ⅜ in. wood structural panel sheathing on one side
- Full-length header to the first stud of each panel
- Minimum header size 3 in. by 11¼ in. net
- Header strapped to the wall to resist uplift (capacity of strap varies based on opening size, wall height, wind speed, and exposure category)
- Continuous concrete foundation with reinforcement

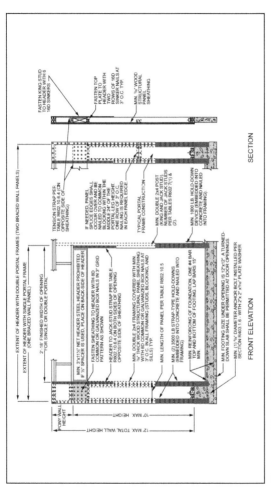

Figure 6.4 Method PFH: portal frame with hold-downs

Source: International Residential Code for One- and Two-Family Dwellings, 2015. Washington, DC: International Code Council, 2014, p. 190.

143

- One ⅝ in. anchor bolt ³⁄₁₆ in. × 2½ in. × 2½ in. plate washer and double bottom plates
- End studs with embedded strap-type hold-down devices rated at 3,500 lb.

Method PFG: Portal Frame at Garage Door Openings in Seismic Design Categories A, B, and C

This method does not require hold-down devices anchored to the foundation but is limited to buildings located in the lower seismic design categories—SDC A, B, and C. This portal frame method also does not get full credit for a 48-in. braced wall panel. The code permits these alternate braced wall panels on either side of garage door openings. Construction must comply with the details of figure 6.5, as follows:

- Panel length must be at least 24 in. and conform to the specifications in table 6.9.
- Wall height, header height, header size, and uplift straps are the same as Method PFH.
- ⁷⁄₁₆ in. wood structural panel sheathing one side
- Two ½ in. anchor bolts with ³⁄₁₆ in. by 2½ in. × 2½ in. plate washers

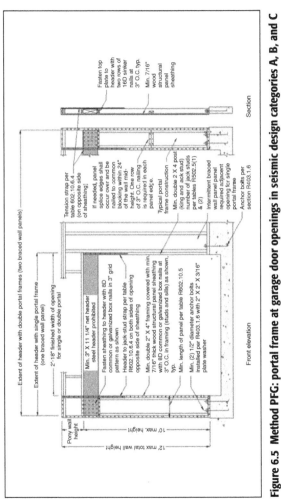

Figure 6.5 Method PFG: portal frame at garage door openings in seismic design categories A, B, and C

Source: International Residential Code for One- and Two-Family Dwellings, 2012. Washington, DC: International Code Council, 2011, p. 179.

145

Method CS-PF: Continuously Sheathed Portal Frame

This continuous sheathing method does not require hold-down devices for anchorage to the foundation but generally requires greater panel lengths than Method PFH. This portal frame method also does not get full credit for a 48 in. braced wall panel. Unlike other portal frame construction, Method CS-PF can be supported by wood floor framing. The code permits a maximum of 4 of these portal frame panels in a single braced wall line. Construction must comply with the details of figure 6.6, which are summarized below:

- Panel length is determined from table 6.9, but must be at least 16 in.
- Wall height, header height, header size, and uplift straps are the same as for Method PFH.
- There must be 7⁄16 in. wood structural panel sheathing on one side.
- When supported by the foundation, two ½ in. anchor bolts with 3⁄16 in. × 2½ in. × 2½ in. plate washers are required.
- Over a raised wood floor, panels must be connected to floor framing with continuous wood structural panel sheathing or framing anchors rated at 670 lbs.

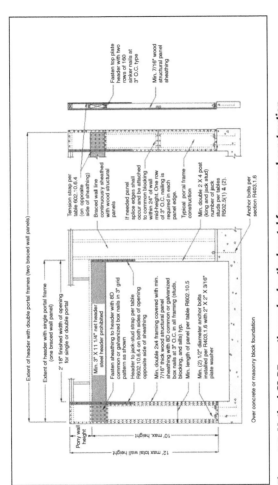

Figure 6.6 Method CS-PF: continuously sheathed portal frame panel construction

Source: International Residential Code for One- and Two-Family Dwellings, 2012. Washington, DC: International Code Council, 2011, p. 180.

Labels within figure:

Fasten top plate header with two rows of 16D sinker nails at 3" O.C. type

Min. 7/16" wood structural panel sheathing

Extent of header with double portal frames (two braced wall panels)

Extent of header with single portal frame (one braced wall panel)

Tension strap per table 602.10.6.4 (on opposite side of sheathing)

Braced wall line continuously sheathed with wood structural panels

If needed panel splice edges shall occur and be attached to common blocking within 24" of wall mid-height. One row of 3" O.C. nailing is required in each panel edge.

Typical portal frame construction

Min. double 2 X 4 post (king and jack stud) number of jack studs per tables R502.5(1) & (2).

Min. 3" X 11 1/4" net header steel header prohibited

Fasten sheathing to header with 8D common or galvanized box nails in 3" grid pattern as shown.

Header to jack-stud strip per table R602.10.6.4 on both sides of opening opposite side of sheathing

Min. double 2x4 framing covered with min. 7/16" thick wood structural panel sheathing with 8D common or galvanized box nails at 3" O.C. in all framing (studs, blocking, and sills) typ.

Min. length of panel per table R602.10.5

Min. (2) 1/2" diameter anchor bolts installed per R403.1.6 with 2" X 2" X 3/16" plate washer

Anchor bolts per section R403.1.6

Over concrete or masonry block foundation

Pony wall height

10' max. height

12' max. total wall height

147

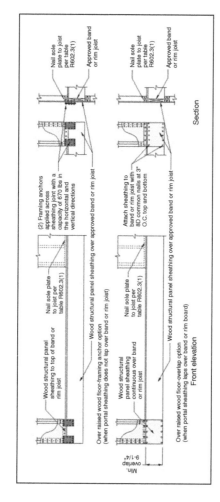

Figure 6.6 Method CS-PF: continuously sheathed portal frame panel construction (*continued***)**

Source: International Residential Code for One- and Two-Family Dwellings, 2012. Washington, DC: International Code Council, 2011, p. 180.

Simplified Wall Bracing

The IRC offers a simplified method of wall bracing that can be used in most parts of the country without requiring complex calculations or adjustment factors. This method is limited to buildings that meet all of the following requirements:

- Maximum three stories above a concrete or masonry foundation or basement wall
- No floor cantilever more than 24 in.
- Maximum wall height of 10 ft.
- Maximum roof eave-to-ridge height of 15 ft.
- Minimum ½ in. gypsum board on the interior side of walls
- Maximum ultimate design wind speed 130 mph and Exposure Category B or C
- SDC A, B, or C for detached one- and two-family dwellings
- SDC A or B for townhomes
- No cripple walls permitted in three-story buildings

Circumscribed Rectangle

When a rectangle is drawn around the building including the attached garage but excluding decks, open porches, and carports, the following requirements apply (fig. 6.7):

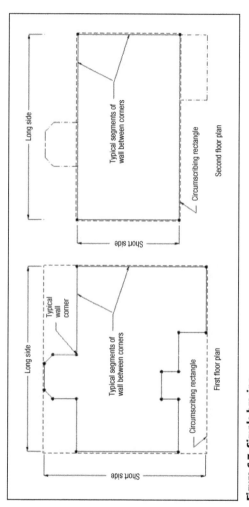

Long side

Typical segments of
wall between corners

Circumscribing rectangle

Second floor plan

Short side

Typical
wall
corner

Long side

Typical segments of
wall between corners

Circumscribing rectangle

First floor plan

Short side

Figure 6.7 Simple bracing

- Maximum 60 ft. dimension on any side
- Maximum ratio of long side to short side of 3:1

Bracing Units

For simplified wall bracing, each section of bracing is called a bracing unit rather than a braced wall panel. Bracing units must

- Be constructed with minimum ⅜ in. wood structural panels or ½ in. structural fiberboard sheathing
- Be at least 3 ft. long when continuous sheathing is of the same material
- Be at least 4 ft. long for intermittent bracing units
- Have no more than 20 ft. of spacing from edge to edge
- Not be inset more than 12 ft. from the corner
- Have at least 1 bracing unit for wall segments longer than 8 ft.

The minimum number of bracing units is determined from table 6.10.

Narrow panels

The code permits using alternative narrow panel construction methods CS-PF, PFH, and PFG as discussed under the conventional wall bracing

Table 6.10 Minimum number of bracing units on each side of the circumscribed rectangle

Ultimate design wind Speed (mph)	Story level	Eave-to-ridge height (ft.)	Minimum number of bracing units on each long side						Minimum number of bracing units on each short side					
			Length of short side (ft.)						Length of long side (ft.)					
			10	20	30	40	50	60	10	20	30	40	50	60
115		10	1	2	2	2	3	3	1	2	2	2	3	3
		10	2	3	3	4	5	6	2	3	3	4	5	6
		10	2	3	4	6	7	8	2	3	4	6	7	8
		15	1	2	3	3	4	4	1	2	3	3	4	4
		15	2	3	4	5	6	7	2	3	4	5	6	7
		15	2	4	5	6	7	9	2	4	5	6	7	9

130

10			15		
1	2	2	2	3	3
2	3	4	3	4	6
2	4	5	3	6	7
3	5	7	4	7	10
3	6	8	4	8	11
4	7	10	6	10	13
1	2	2	2	3	3
2	3	4	3	4	6
2	4	5	3	6	7
3	5	7	4	7	10
3	6	8	4	8	11
4	7	10	6	10	13

Note: For Exposure Category C, multiply bracing units by a factor of 1.20 for a one-story building, 1.30 for a two-story building and 1.40 for a three-story building.

provisions, in addition to Method CS-G, (continuously sheathed wood structural panel adjacent to garage openings). The code requires continuous wood structural panel sheathing on all walls and assigns partial credit in number of bracing units for these narrow panel methods as follows:

- CSG = 0.5 bracing unit
- CS-PF = 0.75 bracing unit
- PFG = 0.75 bracing unit
- PFH = 1.0 bracing unit

Note: Method CS-G is only permitted on one-story garages.

Steel Wall Framing

The code contains prescriptive requirements for cold-formed steel wall framing applicable to buildings not greater than 40 × 60 ft. and up to 3 stories high. Use of these code provisions is further limited to sites with an ultimate design wind speed less than 139 mph and a ground snow load of 70 psf.

Exterior Windows and Doors

Windows and doors located in exterior walls must be installed, anchored, and flashed according to the manufacturer's written installation instructions. The code requires the manufacturer to provide written installation instructions for each window or door.

Structural Insulated Panel (SIP) Wall Construction

The IRC contains prescriptive requirements for SIP wall construction that applies to buildings that do not exceed the following criteria:

- 60 ft. long
- 40 ft. wide
- 2 stories high
- 10 ft. wall height
- Seismic Design Categories A, B, and C
- 70 psf ground snow load
- 155 mph ultimate design wind speed in Exposure B or 140 mph in Exposure C

The code provides tabular values for the minimum thickness of SIP wall construction based on

- Building width
- Roof or story and roof being supported
- Wall height
- Snow load
- Wind speed
- Wind exposure category

7

Wall Covering

This chapter addresses the design and construction of interior and exterior wall coverings. Interior gypsum board (drywall) and plaster must meet minimum thicknesses. The code includes minimum fastener requirements as well. Exterior surfaces must be protected from the weather with appropriate water-resistive barriers, flashing, and finish surfaces such as siding, veneer, or plaster (stucco).

Interior Wall Coverings

Interior wall coverings shall not be installed until the area is protected from the weather. The code contains prescriptive installation requirements for gypsum board (drywall). Minimum fastening and thickness requirements are dependent on the spacing of the framing members and the location and intended application of the gypsum board. The code also contains tables for minimum thicknesses of plaster coats, but refers installers to applicable ASTM standards for material and installation details.

Vapor Retarders

A Class I or Class II vapor retarder must be installed on the interior side of above-grade frame walls in Climate Zones 5, 6, 7, 8, and Marine 4. Sheet polyethylene is an example of a Class I vapor retarder. Kraft-faced fiberglass batt insulation is an example of a Class II vapor retarder.

Gypsum Board Application

Gypsum board can be installed perpendicular or parallel to framing members in accordance with table 7.1, with the following exceptions:

- ⅜ in. must be installed perpendicular to ceiling framing members and is not permitted on ceilings receiving water-based textures or supporting insulation.
- ½ in. must be applied perpendicular to ceiling members spaced 16 in. O.C. when ceilings will receive water-based texture finishes (unless high-strength gypsum board is used).
- ⅝ in. must be applied perpendicular to ceiling members when ceilings will receive water-based texture finishes.

Table 7.1 Minimum thickness and application of gypsum board

| Thickness of gypsum board (in.) | Application | Orientation of gypsum board to framing | Maximum spacing of framing members (in. O.C.) | Maximum spacing of fasteners (in.) | | Size of nails for application to wood framing |
				Nails	Screws	
Application without adhesive						
⅜	Ceiling	Perpendicular	16	7	12	13 gauge, 1¼" long, ¹⁹⁄₆₄" head; 0.098" diameter, 1¼" long, annular-ringed; or 4d cooler nail, 0.080" diameter, 1⅜" long ⁷⁄₃₂" head
	Wall	Either direction	16	8	16	
½	Ceiling	Either direction	16	7	12	13 gauge, 1⅜" long, ¹⁹⁄₆₄" head; 0.098" diameter, 1¼" long, annular-ringed; 5d cooler nail, 0.086" diameter, 1⅝" long, ¹⁵⁄₆₄" head; or gypsum board nail, 0.086" diameter, 1⅝" long, ⁹⁄₃₂" head
	Ceiling	Perpendicular	24	7	12	
	Wall	Either direction	24	8	12	
	Wall	Either direction	16	8	16	

(continued)

Table 7.1 Minimum thickness and application of gypsum board (*continued*)

Thickness of gypsum board (in.)	Application	Orientation of gypsum board to framing	Maximum spacing of framing members (in. O.C.)	Maximum spacing of fasteners (in.)		Size of nails for application to wood framing
				Nails	Screws	
	Application without adhesive (*continued*)					
	Ceiling	Either direction	16	7	12	13 gauge, 1⅝" long, ¹⁹⁄₆₄" head; 0.098" diameter, 1⅜" long, annular-ringed; 6d cooler nail, 0.092" diameter, 1⅞" long, ¼" head; or gypsum board nail, 0.0915" diameter, 1⅞" long, ¹⁹⁄₆₄" head
	Ceiling	Perpendicular	24	7	12	
⅝	Wall	Either direction	24	8	12	
	Wall	Either direction	16	8	16	
	Type X at garage ceiling beneath habitable rooms	Perpendicular	24	6	6	1⅞" long 6d coated nails or equivalent drywall screws

160

Application with adhesive

3⁄8	Ceiling	Perpendicular	16	16	16	Same as above for 3⁄8″ gypsum board
	Wall	Either direction	16	16	24	
	Ceiling	Either direction	16	16	16	
1⁄2 or 5⁄8	Ceiling	Perpendicular	24	12	16	Same as above for 1⁄2″ and 5⁄8″ gypsum board, respectively
	Wall	Either direction	24	16	24	
	Ceiling	Perpendicular	16	16	16	
Two 3⁄8 layers	Wall	Either direction	24	24	24	Base ply nailed as above for 1⁄2″ gypsum board; face ply installed with adhesive

Note: See specific limitations on ceiling applications related to support of insulation, water-based textures, and fire resistance.

- ⅝ in. Type X installed on ceilings of garages to separate them from habitable rooms located above the garage must be perpendicular to the framing members.

Table 7.1 also includes nail and screw fastening requirements. The type of screws to be used depends on the material the gypsum board will be attached to as follows:

- Wood—type W or S screws that penetrate wood at least ⅝ in.
- Cold-formed steel framing—approved screws that comply with the referenced standards and that penetrate steel at least ⅜ in.
- *Structural insulated panels*—approved screws that penetrate wood structural panel sheathing at least ⁷⁄₁₆ in.

Water-Resistant Gypsum Backing Board

Cut or exposed edges must be sealed. Water-resistant gypsum backing board is permitted on ceilings where framing spacing does not exceed 12 in. O.C. for ½ in., or 16 in. O.C. for ⅝ in. It is not allowed

- Over a vapor retarder in a shower or tub compartment
- In areas with direct exposure to water or that are subject to continuous high humidity

- As a backer for ceramic wall tile in tub and shower areas

Ceramic Tile

Ceramic tile surfaces must be installed according to the ANSI standards referenced. The code also requires one of the following materials as a backer for ceramic wall tile in tub and shower areas:

- Fiber cement
- Fiber-mat-reinforced cement
- Glass mat gypsum backer
- Fiber-reinforced gypsum backer

Exterior Wall Coverings

Table 7.2 lists siding and veneer attachment and installation requirements. Wall coverings must provide a weather-resistant exterior wall envelope including the following:

- Flashings at openings, penetrations, and intersections of materials
- Water-resistant barrier behind exterior veneer or siding
- Means of draining water that enters the exterior wall assembly to the exterior
- Protection against condensation in the assembly

Table 7.2 Siding minimum attachment and minimum thickness

Siding material		Nominal thickness (in.)	Joint treatment	Type of supports for the siding material and fasteners						
				Wood or wood structural panel sheathing into stud	Fiberboard sheathing into stud	Gypsum sheathing into stud	Foam plastic sheathing into stud	Direct to studs		Number or spacing of fasteners
Anchored veneer: brick, concrete, masonry or stone		2	Per Section R703.8	Per Section R703.8						
Adhered veneer: concrete, stone or masonry		—	Per Section R703.12	Per Section R703.12						
Fiber-cement siding	Panel siding	5⁄16	Per Section R703.10.1	6d common (2" × 0.113")	6d common (2" × 0.113")	6d common (2" × 0.113")	6d common (2" × 0.113")	4d common (1½" × 0.099")		6" panel edges 12" inter. sup.
	Lap siding	5⁄16	Per Section R703.10.2	6d common (2" × 0.113")	6d common (2" × 0.113")	6d common (2" × 0.113")	6d common (2" × 0.113")	6d common (2" × 0.113") or 11 gage roofing nail		—

Hardboard panel siding		7/16	—	0.120" nail (shank) with 0.225" head	0.120" nail (shank) with 0.225" head	0.120" nail (shank) with 0.225" head	0.120" nail (shank) with 0.225" head	6" panel edges 12" inter. sup.
Hardboard lap siding		7/16	—	0.099" nail (shank) with 0.240" head	0.099" nail (shank) with 0.240" head	0.099" nail (shank) with 0.240" head	0.099" nail (shank) with 0.240" head	Same as stud spacing 2 per bearing
Horizontal aluminum	Without insulation	0.019	Lap	Siding nail 1½" × 0.120"	Siding nail 2" × 0.120"	Siding nail 2" × 0.120"	Not allowed	Same as stud spacing
	Without insulation	0.024	Lap	Siding nail 1½" × 0.120"	Siding nail 2" × 0.120"	Siding nailh 1½" × 0.120"	Not allowed	
	With insulation	0.019	Lap	Siding nail 1½" × 0.120"	Siding nail 2½" × 0.120"	Siding nailh 1½" × 0.120"	Siding nail 1½" × 0.120"	
Insulated vinyl siding		0.035 (vinyl siding layer only)	Lap	0.120 nail (shank) with a 0.313 head or 16 gauge crown	0.120 nail (shank) with a 0.313 head or 16 gauge crown	0.120 nail (shank) with a 0.313 head per Section R703.11.2	Not allowed	16 in. on center or specified by manufacturer instructions, test report or other sections of the code

(continued)

Table 7.2 Siding minimum attachment and minimum thickness (*continued*)

| Siding material | Nominal thickness (in.) | Joint treatment | Type of supports for the siding material and fasteners | | | | | | Number or spacing of fasteners |
			Wood or wood structural panel sheathing into stud	Fiberboard sheathing into stud	Gypsum sheathing into stud	Foam plastic sheathing into stud	Direct to studs	
Particleboard panels	⅜	–	6d box nail (2" × 0.099")	6d box nail (2" × 0.099")	6d box nail (2" × 0.099")	6d box nail (2" × 0.099")	Not allowed	
	½	–	6d box nail (2" × 0.099")	6d box nail (2" × 0.099")	6d box nail (2" × 0.099")	6d box nail (2" × 0.099")	6d box nail (2" × 0.099")	6" panel edges 12" inter. sup.
	⅝	–	6d box nail (2" × 0.099")	8d box nail (2½" × 0.113")	8d box nail (2½" × 0.113")	6d box nail (2" × 0.099")	6d box nail (2" × 0.099")	
Polypropylene siding	Not applicable	Lap	Section 703.14.1	Section 703.14.1	Section 703.14.1	Section 703.14.1	Not allowed	As specified by the manufacturer instructions, test report or other sections of this code

Steel		29 ga.	Lap	Siding nail (1¾" × 0.113") Staple–1¾"	Siding nail (2¾" × 0.113") Staple–2½"	Siding nail (2½" × 0.113") Staple–2¼"	Siding nail (1¾" × 0.113") Staple–1¾"	Not allowed	Same as stud spacing
Vinyl siding		0.035	Lap	0.120" nail (shank) with a 0.313" head or 16 gauge staple with ⅜–½" crown	0.120" nail (shank) with a 0.313" head or 16 gauge staple with ⅜–½" crown	0.120" nail (shank) with a 0.313" head or 16 gauge staple with ⅜–½" crown	0.120 nail (shank) with a 0.313 head per Section R703.11.2	Not allowed	16 in. on center or specified by the manufacturer instructions or test report
Wood siding	Wood rustic, drop	⅜ Min	Lap	6d box or siding nail (2" × 0.099")	6d box or siding nail (2" × 0.099")	6d box or siding nail (2" × 0.099")	6d box or siding nail (2" × 0.099")	8d box or siding nail (2½" × 0.113") Staple–2"	Face nailing up to 6" widths, 1 nail per bearing; 8" widths and over, 2 nails per bearing
	Shiplap	19/32 Average	Lap						
	Bevel	7/16							
	Butt tip	3/16	Lap						
Wood structural panel ANSI/APA PRP-210 siding (exterior grade)		⅜–½	–	2" × 0.099" siding nail	2½" × 0.113" siding nail	2½" × 0.113" siding nail	2½" × 0.113" siding nail	2" × 0.099" siding nail	6" panel edges 12" inter. sup.
Wood structural panel lap siding		⅜–½	–	2" × 0.099" siding nail	2½" × 0.113" siding nail	2½" × 0.113" siding nail	2½" × 0.113" siding nail	2" × 0.099" siding nail	8" along bottom edge

Alternative assemblies are permitted if they have been proven to resist wind-driven rain. Additional covering is not required over masonry and concrete walls that are properly flashed.

Water-Resistive Barrier

One layer of No. 15 asphalt felt or other approved water-resistive barrier, such as an approved house wrap, is required over studs or sheathing of all exterior walls. Water-resistive barriers must be applied as follows:

- Horizontally
- The upper layers must be lapped over the lower layers at least 2 in.
- Vertical joints must be lapped at least 6 in.
  Note: *Water-resistive barrier is not required for detached accessory buildings.*

Flashing Requirements Applicable to All Exterior Wall Coverings

To prevent water entering behind exterior wall coverings and penetrating the wall assembly, the code requires corrosion-resistant flashing at the following locations:

- Exterior window and door openings
- The intersection of masonry construction with frame or stucco walls

- Under and at the ends of masonry, wood, or metal copings and sills
- Continuously above all projecting wood trim
- Where exterior porches, decks, or stairs attach to a wall or floor
- At wall and roof intersections

Exterior Insulation and Finish Systems (EIFS)

Installation of EIFS must comply with the referenced ASTM standards and the manufacturer's instructions to provide the code-prescribed weather-resistant exterior envelope. The code permits EIFS without drainage over masonry and concrete construction only. For all other EIFS installations, including those over wood-frame construction, the IRC requires EIFS with drainage. To further protect the integrity of the system, the code also stipulates the following:

- No face nailing of decorative trim through EIFS
- Termination at least 6 in. above the finished ground level

Siding

The code includes installation requirements for both panel and horizontal lapped siding made of wood, hardboard, wood structural panel, fiber cement, and vinyl. Installation must follow the manufacturer's requirements and the code provisions.

Wood, Hardboard, and Wood Structural Panel Siding

For panel siding, horizontal joints must occur over solid blocking or wood structural panel sheathing and comply with one of the following:

- Lapped at least 1 in.
- Shiplapped
- Flashed with z flashing

Vertical joints must occur over framing members or wood structural panel sheathing and be either shiplapped or covered with a batten.

For horizontal lap siding, laps must comply with the manufacturer's recommendations or, if there are no recommendations, the lap must be at least 1 in. (or ½ in. if *rabbeted*). End joints must be made weather tight with one of the following methods:

- Caulked
- Covered with a batten
- Sealed and installed over a strip of flashing

Fiber Cement Siding

Fiber cement siding is manufactured of portland cement, sand, wood fiber, and specialty additives. For vertical panel applications, vertical and horizontal joints must

- Occur over framing members
- Be sealed, covered with battens or flashed

For horizontal lap siding applications, the code requires at least a 1¼ in. lap. End joints require one of the following treatments:

- Sealed with caulking
- Covered with an H-section joint cover
- Located over a strip of flashing
- Tongue and groove

Vinyl Siding

The IRC references the manufacturer's instructions for installing vinyl siding, soffit, and accessories. Each soffit panel must be fastened to framing or supporting components such as:

- Nailing strips
- Fascia or sub-fascia

When vinyl siding is installed over foam plastic sheathing that is not backed with structural sheathing, the code sets specific criteria for resisting design wind pressure. For locations with an ultimate design wind speed not greater than 115 mph that fall under Exposure Category B, the minimum fastener requirements are:

- 1¼ in. penetration into wood framing
- 0.120-in. diameter nail
- 0.313-in. diameter head
- Nails spaced 16 in. O.C.

Vinyl siding installation in locations with a greater wind speed or exposure category must be

- Adjusted according to the design wind pressures or
- Comply with the manufacturer's specifications for the applicable wind pressure rating

Wood Shakes and Shingles

As with other exterior wall coverings, wood shakes and shingles require an approved water-resistive barrier, such as felt or house wrap. Other installation requirements are as follows:

- Shakes and shingles are attached to wood-based sheathing or furring strips.
- Spacing between shingles is ⅛ to ¼ in.
- Spacing between shakes is ⅜ to ½ in.
- Each shingle or shake requires two hot-dipped, zinc-coated steel, stainless steel, or aluminum fasteners.
- Fasteners shall penetrate sheathing or furring strips at least ½ in.
- Bottom courses must be doubled.
- The offset of joints in adjacent courses must be at least 1½ in.

Stone and Masonry Veneer

In general, stone and masonry veneers are limited to the first story and are not greater than 5 in. thick, according to the prescriptive provisions of the IRC. Exceptions allow veneers up to 3 stories and 30 ft. above noncombustible foundations (plus 8 ft. for gables) for wood frame construction depending on the SDC, the nominal thickness and weight of the veneer, and the use of the building (tables 7.3 and 7.4).

Support

Masonry veneer typically is supported by a continuous concrete or masonry foundation. In seismic design categories A, B, and C, the code permits the following methods of support:

Table 7.3 Stone or masonry veneer limitations and requirements, wood or steel framing, seismic design categories A, B, and C

Seismic design category	Wood or steel framing	Number of stories	Maximum height of veneer above noncombustible foundation (ft.)	Maximum nominal thickness of veneer (in.)	Maximum weight of veneer (psf)
A, B, or C	Steel	1 or 2	30	5	50
	Wood	1, 2, or 3	30	5	50

Note: An additional 8 ft. of height is permitted for gable end walls.

Table 7.4 Stone or masonry veneer limitations and requirements, one- and two-family detached dwellings, wood framing, seismic design categories D_0, D_1, and D_2

Seismic design category	Number of wood framed stories	Maximum height of veneer above noncombustible foundation (ft.)	Maximum nominal thickness of veneer (in.)	Maximum weight of veneer (psf)
D_0	1 or 2	20	4	40
	3	30	4	40
D_1	1 or 2	20	4	40
	3	20	4	40
D_2	1 or 2	20	3	30

Note: An additional 8 ft. of height is permitted for gable end walls.

■ Directly on wood frame or steel frame construction when designed to limit deflection to 1/600 of the span of the supporting members
■ On a steel angle support measuring at least 6 × 4 × 5/16 in. and attached to wall construction as follows:
  • Long leg of angle placed vertically
  • Attached to double 2 × 4 studs with two 7/16 × 4 in. lag screws at 16 in. O.C.
  • Bearing of at least 2/3 of veneer width

- Flashing and weep holes above the angle
- Maximum 12 ft., 8 in. veneer height

Lintels

To support veneer above openings, the code requires noncombustible lintels with a bearing of at least 4 in. at each end. Table 7.5 provides spans for both steel angle and reinforced masonry lintels based on the number of stories above the lintel. Steel angle lintels must be installed with the long leg of the

Table 7.5 Allowable spans for lintels supporting masonry veneer

Size of steel angle (in.)	No story above (ft.-in.)	One story above (ft.-in.)	Two stories above (ft.-in.)	No. of ½″ or equivalent reinforcing bars
3 × 3 × ¼	6-0	4-6	3-0	1
4 × 3 × ¼	8-0	6-0	4-6	1
5 × 3½ × ⁵⁄₁₆	10-0	8-0	6-0	2
6 × 3½ × ⁵⁄₁₆	14-0	9-6	7-0	2
2–6 × 3½ × ⁵⁄₁₆	20-0	12-0	9-6	4

Note: Long leg of the angle shall be placed in a vertical position.
Depth of reinforced lintels shall not be less than 8 in., and all cells of hollow masonry lintels shall be grouted solid.
Reinforcing bars shall extend not less than 8 in. into the support.
Steel members indicated are adequate typical examples; other steel members meeting structural design requirements may be used.

angle situated vertically. They must be protected from corrosion in one of the following ways:

- Shop-coated with rust-inhibitive paint
- Made of corrosion-resistant steel
- Treated with corrosion-resistant coating

Veneer Anchored with Corrosion-Resistant Metal Ties
The IRC prescribes the type, size, and location of corrosion-resistant metal ties to anchor the veneer to the supporting structure as follows:

- Minimum No. 9 U.S. gage strand wire ties
- Minimum No. 22 U.S. gage × ⅞ in. corrugated sheet metal ties
- Maximum spacing 32 in. O.C. horizontally and 24 in. O.C. vertically
- Maximum supported area 2.67 sq. ft. per tie
- Additional ties around openings
- Minimum 1½ in. tie embedment
- Minimum ⅝ in. mortar cover to outside face

Note: *In SDC D_0, D_1, D_2, or for townhomes in SDC C each tie shall support no more than 2 sq. ft. of wall area.*

Veneer Details
To provide an adequate weather-resistant exterior wall envelope, exterior stone and masonry veneer require an air space and water-resistant components

to prevent moisture from entering into the wall assembly (figs. 7.1a and 7.1b). The IRC requires the following:

- An approved water-resistant barrier over the wall sheathing and behind the veneer
- An air space of nominal 1 in. between the veneer and the sheathing when corrugated sheet metal ties are used
- An air space of nominal 1 in. to 4½ in. between the veneer and the sheathing when strand wire ties are used
- Flashing below the first course of masonry veneer above grade and above other points of support
- Weep holes of at least ³⁄₁₆ in. diameter and spaced no more than 33 in. O.C. immediately above the flashing

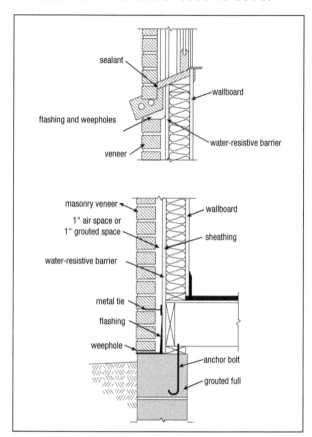

Figure 7.1a Masonry veneer details at foundation and sill

Source: International Residential Code for One- and Two-Family Dwellings, 2006. Washington, DC: International Code Council Inc., 2006, p. 236.

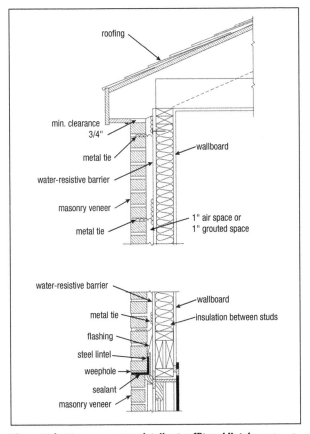

roofing

min. clearance 3/4"

metal tie

water-resistive barrier

masonry veneer

metal tie

wallboard

1" air space or 1" grouted space

water-resistive barrier

metal tie

flashing

steel lintel

weephole

sealant

masonry veneer

wallboard

insulation between studs

Figure 7.1b Masonry veneer details at soffit and lintel (*continued*)

Source: International Residential Code for One- and Two-Family Dwellings, 2006. Washington, DC: International Code Council Inc., 2006, p. 237.

8

Roof Framing

This chapter addresses conventional light-frame roof ceiling construction of both wood and cold-formed steel materials. Maximum ceiling joist spans are for attics without, or with limited, storage (tables 8.1–8.2). (Habitable attics and attics with fixed-stair access require joists sized as floor joists.) The prescriptive tables setting maximum spans for rafters are based on the snow load of the geographic area and whether the rafter also supports a ceiling (tables 8.3–8.8). The tables assume a roof live load of 20 psf for regions without snow loads. This chapter also covers attic access and ventilation.

Wood Roof Framing Details

The conventional roof framing provisions in the code apply to roofs with a slope of not less than 3:12. The code prescribes the following framing details:

Table 8.1 Ceiling joist spans for common lumber species, No. 2 grade

(uninhabitable attics without storage, live load = 10 psf, L/240)

Ceiling joist spacing (in.)	Species	Dead load = 5 psf			
		2 × 4	2 × 6	2 × 8	2 × 10
		Maximum ceiling joist spans			
		(ft.-in.)	(ft.-in.)	(ft.-in.)	(ft.-in.)
12	Douglas fir-larch	12-5	19-6	25-8	> 26-0
	Hem-fir	11-7	18-2	24-0	> 26-0
	Southern pine	11-10	18-8	24-7	> 26-0
	Spruce-pine-fir	11-10	18-8	24-7	> 26-0
16	Douglas fir-larch	11-3	17-8	23-4	> 26-0
	Hem-fir	10-6	16-6	21-9	> 26-0
	Southern pine	10-9	16-11	21-7	25-7
	Spruce-pine-fir	10-9	16-11	22-4	> 26-0
19.2	Douglas fir-larch	10-7	16-8	21-4	26-0
	Hem-fir	9-11	15-7	20-6	25-3
	Southern pine	10-2	15-7	19-8	23-5
	Spruce-pine-fir	10-2	15-11	21-0	25-8
24	Douglas fir-larch	9-10	15-0	19-1	23-3
	Hem-fir	9-2	14-5	18-6	22-7
	Southern pine	9-3	13-11	17-7	20-11
	Spruce-pine-fir	9-5	14-9	18-9	22-11

Table 8.2 Ceiling joist spans for common lumber species, No. 2 grade

(uninhabitable attics with limited storage, live load = 20 psf, L/240)

Ceiling joist spacing (in.)	Species	Dead load = 10 psf			
		2 × 4	2 × 6	2 × 8	2 × 10
		Maximum ceiling joist spans			
		(ft.-in.)	(ft.-in.)	(ft.-in.)	(ft.-in.)
12	Douglas fir-larch	9-10	15-0	19-1	23-3
	Hem-fir	9-2	14-5	18-6	22-7
	Southern pine	9-3	13-11	17-7	20-11
	Spruce-pine-fir	9-5	14-9	18-9	22-11
16	Douglas fir-larch	8-11	13-0	16-6	20-2
	Hem-fir	8-4	12-8	16-0	19-7
	Southern pine	8-0	12-0	15-3	18-1
	Spruce-pine-fir	8-7	12-10	16-3	19-10
19.2	Douglas fir-larch	8-2	11-11	15-1	18-5
	Hem-fir	7-10	11-7	14-8	17-10
	Southern pine	7-4	11-0	13-11	16-6
	Spruce-pine-fir	8-0	11-9	14-10	18-2
24	Douglas fir-larch	7-3	10-8	13-6	16-5
	Hem-fir	7-1	10-4	13-1	16-0
	Southern pine	6-7	9-10	12-6	14-9
	Spruce-pine-fir	7-2	10-6	13-3	16-3

Table 8.3 Rafter spans for common lumber species, No. 2 grade

(roof live load = 20 psf, ceiling not attached to rafters, L/180,
dead load = 10 psf)

Rafter spacing (in.)	Species	Dead load = 10 psf				
		2 × 4	2 × 6	2 × 8	2 × 10	2 × 12
		Maximum rafter spans				
		(ft.-in.)	(ft.-in.)	(ft.-in.)	(ft.-in.)	(ft.-in.)
12	Douglas fir-larch	10-10	16-10	21-4	26-0	> 26-0
	Hem-fir	10-1	15-11	20-8	25-3	> 26-0
	Southern pine	10-4	15-7	19-8	23-5	> 26-0
	Spruce-pine-fir	10-4	16-3	21-0	25-8	> 26-0
16	Douglas fir-larch	9-10	14-7	18-5	22-6	26-0
	Hem-fir	9-2	14-2	17-11	21-11	25-5
	Southern pine	9-0	13-6	17-1	20-3	23-10
	Spruce-pine-fir	9-5	14-4	18-2	22-3	25-9
19.2	Douglas fir-larch	9-1	13-3	16-10	20-7	23-10
	Hem-fir	8-8	12-11	16-4	20-0	23-2
	Southern pine	8-2	12-3	15-7	18-6	21-9
	Spruce-pine-fir	8-10	13-1	16-7	20-3	23-6
24	Douglas fir-larch	8-2	11-11	15-1	18-5	21-4
	Hem-fir	7-11	11-7	14-8	17-10	20-9
	Southern pine	7-4	11-0	10-11	16-6	19-6
	Spruce-pine-fir	8-0	11-9	14-10	18-2	21-0

Table 8.4 Rafter spans for common lumber species, No. 2 grade

(roof live load = 20 psf, ceiling attached to rafters, L/240,
dead load = 10 psf)

Rafter spacing (in.)	Species	Dead load = 10 psf				
		2 × 4	**2 × 6**	**2 × 8**	**2 × 10**	**2 × 12**
		Maximum rafter spans				
		(ft.-in.)	(ft.-in.)	(ft.-in.)	(ft.-in.)	(ft.-in.)
12	Douglas fir-larch	9-10	15-6	20-5	26-0	> 26-0
	Hem-fir	9-2	14-5	19-0	24-3	> 26-0
	Southern pine	9-5	14-9	19-6	23-5	> 26-0
	Spruce-pine-fir	9-5	14-9	19-6	24-10	> 26-0
16	Douglas fir-larch	8-11	14-1	18-5	22-6	26-0
	Hem-fir	8-4	13-1	17-3	21-11	25-5
	Southern pine	8-7	13-5	17-1	20-3	23-10
	Spruce-pine-fir	8-7	13-5	17-9	22-3	25-9
19.2	Douglas fir-larch	8-5	13-3	16-10	20-7	23-10
	Hem-fir	7-10	12-4	16-3	20-0	23-2
	Southern pine	8-1	12-3	15-7	18-6	21-9
	Spruce-pine-fir	8-1	12-8	16-7	20-3	23-6
24	Douglas fir-larch	7-10	11-11	15-1	18-5	21-4
	Hem-fir	7-3	11-5	14-8	17-10	20-9
	Southern pine	7-4	11-0	13-11	16-6	19-6
	Spruce-pine-fir	7-6	11-9	14-10	18-2	21-0

Table 8.5 Rafter spans for common lumber species, No. 2 grade

(ground snow load = 30 psf, ceiling not attached to rafters, L/180, dead load = 10 psf)

Rafter spacing (in.)	Species	Dead Load = 10 psf				
		2 × 4	2 × 6	2 × 8	2 × 10	2 × 12
		Maximum rafter spans				
		(ft.-in.)	(ft.-in.)	(ft.-in.)	(ft.-in.)	(ft.-in.)
12	Douglas fir-larch	9-6	14-0	17-8	21-7	25-1
	Hem-fir	8-10	13-7	17-2	21-0	24-4
	Southern pine	8-7	12-11	16-4	19-5	22-10
	Spruce-pine-fir	9-1	13-9	17-5	21-4	24-8
16	Douglas fir-larch	8-3	12-1	15-4	18-9	21-8
	Hem-fir	8-0	11-9	14-11	18-2	21-1
	Southern pine	7-6	11-2	14-2	16-10	19-10
	Spruce-pine-fir	8-2	11-11	15-1	18-5	21-5
19.2	Douglas fir-larch	7-7	11-0	14-0	17-1	19-10
	Hem-fir	7-4	10-9	13-7	16-7	19-3
	Southern pine	6-10	10-2	12-11	15-4	18-1
	Spruce-pine-fir	7-5	10-11	13-9	16-10	19-6
24	Douglas fir-larch	6-9	9-10	12-6	15-3	17-9
	Hem-fir	6-7	9-7	12-2	14-10	17-3
	Southern pine	6-1	9-2	11-7	13-9	16-2
	Spruce-pine-fir	6-8	9-9	12-4	15-1	17-6

Table 8.6 Rafter spans for common lumber species, No. 2 grade
(ground snow load = 50 psf, ceiling not attached to rafters, L/180, dead load = 10 psf)

Rafter spacing (in.)	Species	Dead load = 10 psf				
		2 × 4	2 × 6	2 × 8	2 × 10	2 × 12
		Maximum rafter spans				
		(ft.-in.)	(ft.-in.)	(ft.-in.)	(ft.-in.)	(ft.-in.)
12	Douglas fir-larch	7-10	11-5	14-5	17-8	20-5
	Hem-fir	7-5	11-1	14-0	17-2	19-11
	Southern pine	7-0	10-6	13-4	15-10	18-8
	Spruce-pine-fir	7-8	11-3	14-3	17-5	15-2
16	Douglas fir-larch	6-9	9-10	12-6	15-3	17-9
	Hem-fir	6-7	9-7	12-2	14-10	17-3
	Southern pine	6-1	9-2	11-7	13-9	16-2
	Spruce-pine-fir	6-8	9-9	12-4	15-1	17-6
19.2	Douglas fir-larch	6-2	9-0	11-5	13-11	16-2
	Hem-fir	6-0	8-9	11-1	13-7	15-9
	Southern pine	5-7	8-4	10-7	12-6	14-9
	Spruce-pine-fir	6-1	8-11	11-3	13-9	15-11
24	Douglas fir-larch	5-6	8-1	10-3	12-6	14-6
	Hem-fir	5-4	7-10	9-11	12-1	14-1
	Southern pine	5-0	7-5	9-5	11-3	13-2
	Spruce-pine-fir	5-5	7-11	10-1	12-4	14-3

Table 8.7 Rafter spans for common lumber species, No. 2 grade

(ground snow load = 30 psf, ceiling attached to rafters, L/240, dead load = 10 psf)

Rafter spacing (in.)	Species	Dead load = 10 psf				
		2 × 4	2 × 6	2 × 8	2 × 10	2 × 12
		Maximum rafter spans				
		(ft.-in.)	(ft.-in.)	(ft.-in.)	(ft.-in.)	(ft.-in.)
12	Douglas fir-larch	8-7	13-6	17-8	21-7	25-1
	Hem-fir	8-0	12-7	16-7	21-0	24-4
	Southern pine	8-3	12-11	16-4	19-5	22-10
	Spruce-pine-fir	8-3	12-11	17-0	21-4	24-8
16	Douglas fir-larch	7-10	12-1	15-4	18-9	21-8
	Hem-fir	7-3	11-5	14-11	18-2	21-1
	Southern pine	7-6	11-2	14-2	16-10	19-10
	Spruce-pine-fir	7-6	11-9	15-1	18-5	21-5
19.2	Douglas fir-larch	7-4	11-0	14-0	17-1	19-10
	Hem-fir	6-10	10-9	13-7	16-7	19-3
	Southern pine	6-10	10-2	12-11	15-4	18-1
	Spruce-pine-fir	7-0	10-11	13-9	16-10	19-6
24	Douglas fir-larch	6-9	9-10	12-6	15-3	17-9
	Hem-fir	6-4	9-7	12-2	14-10	17-3
	Southern pine	6-1	9-2	11-7	13-9	16-2
	Spruce-pine-fir	6-6	9-9	12-4	15-1	17-6

Table 8.8 Rafter spans for common lumber species, No. 2 grade

(ground snow load = 50 psf, ceiling attached to rafters, L/240,
dead load = 10 psf)

Rafter spacing (in.)	Species	Dead load = 10 psf				
		2 × 4	2 × 6	2 × 8	2 × 10	2 × 12
		Maximum rafter spans				
		(ft.-in.)	(ft.-in.)	(ft.-in.)	(ft.-in.)	(ft.-in.)
12	Douglas fir-larch	7-3	11-5	14-5	17-8	20-5
	Hem-fir	6-9	10-8	14-0	17-2	19-11
	Southern pine	6-11	10-6	13-4	15-10	18-8
	Spruce-pine-fir	6-11	10-11	14-3	17-5	20-2
16	Douglas fir-larch	6-7	9-10	12-6	15-3	17-9
	Hem-fir	6-2	9-7	12-2	14-10	17-3
	Southern pine	6-1	9-2	11-7	13-9	16-2
	Spruce-pine-fir	6-4	9-9	12-4	15-1	17-6
19.2	Douglas fir-larch	6-2	9-0	11-5	13-11	16-2
	Hem-fir	5-9	8-9	11-1	13-7	15-9
	Southern pine	5-7	8-4	10-7	12-6	14-9
	Spruce-pine-fir	5-11	8-11	11-3	13-9	15-11
24	Douglas fir-larch	5-6	8-1	10-3	12-6	14-6
	Hem-fir	5-4	7-10	9-11	12-1	14-1
	Southern pine	5-0	7-5	9-5	11-3	13-2
	Spruce-pine-fir	5-5	7-11	10-1	12-4	14-3

- Typically rafters are framed to a ridge board, but gusset plate ties also are permitted.
- The ridge board must be at least 1 in. nominal thickness and at least as deep as the cut end of the rafter.
- Valley and hip rafters must be at least 2 in. nominal thickness, adequately supported, and at least as deep as the cut end of the rafter.

Note: For roof slopes of less than 3:12, structural members that support rafters and ceiling joists (such as hips and valleys) must be designed the same as beams.

Headers and Trimmers

Header and trimmer requirements for roof and ceiling openings are as follows:

- Headers longer than 4 ft. must be at least double members.
- Trimmers supporting headers longer than 4 ft. must be at least double members.
- Headers and tail joists require hangers or other methods of support at bearing ends.

Framing Connections

The IRC prescribes minimum framing connections to resist the outward thrust of rafters at the wall top

plate and to ensure adequate transfer of all loads to
the supporting structure.

- Ceiling joists and rafters must be nailed together.
- Rafters must be fastened to the top wall plate.
- Ceiling joists must provide a continuous tie across
  the building at the top plate.
- Minimum 2 × 4 rafter ties must be installed if
  ceiling joists are not connected to the rafters at
  the plate or if they are installed perpendicular to
  the rafters.
- A ridge must be supported by a bearing wall or
  a girder or it must be designed as a beam if there
  are no ceiling joists or rafter ties.
- Rafters and ceiling joists require 1½ in. of bearing
  on wood or metal and 3 in. on masonry.
- Connections must provide a continuous load path
  from the roof to the foundation.

Cutting and Notching

Notches in lumber ceiling joists, rafters, and beams
are limited as follows (fig. 8.1):

- Depth less than or equal to ⅙ the depth of the
  member (no more than ¼ at ends)
- Length no longer than ⅓ the depth of the
  member
- Not located in the middle ⅓ of the span

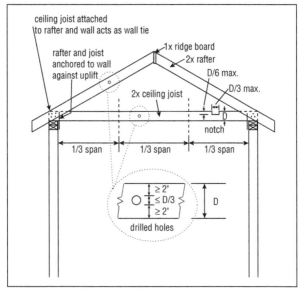

Figure 8.1 Cutting, notching, and drilling

Holes in ceiling joists, rafters, and beams shall be
- Less than or equal in diameter to ⅓ the depth of the member
- At least 2 in. from the top and bottom of the member
- At least 2 in. away from any other hole or notch

Cuts, notches, and holes in engineered wood products (trusses, structural composite lumber, structural

glue-laminated members, or I-joists) are prohibited unless they are
- Permitted by the manufacturer's recommendations
- Part of a specific designed alteration by a registered design professional

Wood Trusses

Wood trusses are to be designed according to accepted engineering practice and referenced standards. Trusses may not be altered without the approval of a registered design professional.
- Truss design drawings are to be
  - Submitted to the building official and approved prior to installation
  - Delivered to the jobsite with the trusses
- Trusses must be braced in accordance with the truss design drawings.

Roof Tie-Down

Structures must have roof-to-wall connections that will resist wind uplift forces and provide a complete load path. The IRC provides prescriptive values for uplift resistance based on building width, wind speed, exposure category, and roof pitch (table 8.9). The type of connection for fastening the rafters or

Table 8.9 Rafter or truss uplift connection forces from wind (pounds per connection)

Rafter or truss spacing	Roof span (ft.)	Exposure B			
		Ultimate design wind speed V$_{ult}$ (mph)			
		115		130	
		Roof pitch		Roof pitch	
		< 5:12	≥ 5:12	< 5:12	≥ 5:12
24 in. O.C.	12	118	106	190	176
	18	148	132	244	224
	24	178	158	298	274
	28	198	176	334	306
	32	218	194	370	340
	36	240	212	406	372
	42	270	240	460	422
	48	302	268	516	472

Rafter or truss spacing	Roof Span (ft.)	Exposure C			
		Ultimate design wind speed V$_{ult}$ (mph)			
		115		130	
		Roof Pitch		Roof Pitch	
		< 5:12	≥ 5:12	< 5:12	≥ 5:12
24 in. O.C.	12	220	204	322	302
	18	282	262	416	390
	24	346	320	512	478
	28	390	358	578	538
	32	432	398	642	598
	36	474	438	706	658
	42	538	496	804	750
	48	604	556	900	840

roof trusses to the top plate is based on the tabular value as follows:

- ☐ ≤ 200 lbs.—toenailed connection permitted (table 6.1) using
  - three 16d box nails or three 10d common nails
  - two toe nails on one side and one toe nail on the reverse side
- ☐ > 200 lbs.—manufactured connector rated to meet or exceed the tabular value

Note: *As an alternative to table 8.9, the builder can use the uplift value specified in the truss design drawings for roof truss-to-wall connection.*

Steel Roof Framing

The code contains prescriptive requirements for cold-formed steel roof framing for buildings not greater than 40 × 60 ft., 3 stories high, and with roof slopes of 3:12 to 12:12. These code provisions are limited to sites with an ultimate design wind speed of less than 139 mph and a ground snow load of 70 psf.

Roof Ventilation

The code requires cross-ventilation for each attic or enclosed roof space as follows:

- Ventilation openings must be covered with corrosion-resistant wire mesh, with $\frac{1}{16}$–$\frac{1}{4}$ in. openings.
- Total net free ventilating area must be at least $\frac{1}{150}$ of the area of the space.
- Free ventilating area of at least $\frac{1}{300}$ of the area of the space is permitted under either of the following conditions:
  - 40%–50% of the required ventilating area is in the upper portion of the space and eave or cornice vents provide the balance of ventilating area.
  - In Climate Zones 6, 7, and 8, a Class I or Class II vapor retarder is installed on the warm-in-winter side of the ceiling.
- Insulation shall not block the free flow of air.
- For ventilated spaces, the code requires at least 1 in. of air space between the insulation and the roof sheathing.

Unvented Attic

In unvented attics, the roof above the attic space forms the thermal envelope. The IRC requires installation of *air-impermeable insulation* (typically rigid board insulation) directly under or directly above the structural roof sheathing, with no insulation or vapor barrier installed below the unvented attic area. Air permeable insulation may be applied directly below the air-impermeable insulation, provided this application satisfies minimum rigid board or air-impermeable insulation *R-value* based on climate zone (table 8.10).

Note: The same principles apply to unvented rafter spaces such as occur in cathedral ceilings.

Attic Access

In wood frame construction, attic areas that exceed 30 sq. ft. and are at least 30 in. high require access as follows:

Table 8.10 Insulation for condensation control above unvented attics

Climate zone	Minimum rigid board or air-impermeable insulation R-value
2B and 3B tile roof only	0 (none required)
1, 2A, 2B, 3A, 3B, 3C	R-5
4C	R-10
4A, 4B	R-15
5	R-20
6	R-25
7	R-30
8	R-35

- The rough opening must be at least 22 × 30 in.
- The opening must be in a hallway or other readily accessible location.
- There must be at least 30 in. headroom above the access.

9

Roof Finishing

Roofing design, materials, construction, and assemblies must provide protection from the weather. Roof coverings must be installed according to the code and the manufacturer's instructions. This chapter addresses roofing underlayment, ice barrier, flashing, and asphalt and wood shingles. It also discusses limitations on reroofing.

Flashing Locations and Materials

Flashing must be corrosion-resistant metal at least 0.019 in. (No. 26 galvanized sheet) or other approved material. It must be installed at the following locations:

- At wall and roof intersections
- Where slope or direction changes
- Around roof openings

Note: The code requires installation of a flashing to divert the water away from where the eave of a sloped roof intersects a vertical sidewall.

Ice Barrier

In areas with a history of water damage to structures from ice dams at roof eaves, asphalt shingles, metal roof shingles, mineral-surfaced roll roofing, slate and slate-type shingles, wood shingles and wood shakes require an ice barrier as follows:

- Self-adhering polymer modified bitumen sheet or two layers of cemented underlayment
- Barrier extending from eave to at least 24 in. inside the exterior wall line of the building

Note: *The ice barrier is not required for unheated detached accessory structures.*

Asphalt Shingles

Asphalt shingles must be installed according to the manufacturer's instructions, using special methods of fastening for slopes greater than 21:12. The minimum roof slope is 2:12.

Fasteners

Fasteners must be galvanized steel, stainless steel, aluminum, or copper roofing nails that meet the following minimum size requirements:

- 12 gage (0.105 in.) shank
- 3/8 in. diameter head

▪ Penetrate at least ¾ in. into roof sheathing or
  through sheathing

Underlayment

Approved asphalt-saturated organic felt or other
approved material must be installed under shingles.
The underlayment lap requirements vary by roof
slope and are as follows:

▪ Slopes of at least 2:12 and less than 4:12 require
  two layers with 19 in. overlaps.
▪ Slopes of 4:12 or greater require a single layer
  with a horizontal lap of at least 2 in.
▪ End laps are offset at least 6 ft.

Valleys

Approved valley linings are to be installed according
to the manufacturer's installation instructions and
must meet one of the following requirements:

▪ Metal valley lining at least 24 in. wide for open or
  closed valley
▪ Two plies of mineral surfaced roll roofing (18 in.
  bottom, 36 in. top layer) for open or closed valley

Other valley linings for closed valleys (covered
with shingles) may include:

▪ One ply of approved smooth roll roofing at least
  36 in. wide
▪ Self-adhering polymer-modified bitumen sheet

Sidewall Flashing

Where the roof joins a sidewall, the IRC requires step flashings or a continuous flashing measuring at least 4 in. wide by 4 in. high. The flashing at the lower end must turn out to direct water away from the wall.

Drip Edge

The code requires a drip edge at eaves and gables (rakes) of shingle roofs. The drip edge must be installed as follows:

- Minimum lap of 2 in.
- Minimum 0.25 in. extension past roof sheathing
- Minimum 2 in. extension onto roof sheathing
- Maximum 12 in. spacing of fasteners (roofing nails)
- Underlayment over drip edge at eaves
- Underlayment under drip edge at gables (rakes)

Crickets and Saddles

Any chimney penetration more than 30 in. wide requires a *cricket* or saddle using the same material as the roof covering or sheet metal.

Wood Shingles and Wood Shakes

The IRC includes requirements specific to wood shingles and shakes used for roofing. In addition to material and grading requirements, the code provides installation details related to slope, decking, underlayment, laps, exposure, and fastening.

Application

- Spacing between shingles is ¼–⅜ in.
- Spacing between shakes is ⅜–⅝ in.
- Side laps are offset at least 2½ in. on adjacent courses.
- Each shake or shingle gets two corrosion-resistant fasteners.
- Fasteners are to penetrate sheathing at least ½ in.
- Shingle fasteners must be no farther than ¾ in. from the edge and no higher than 1 in. above the exposure line.
- Shake fasteners must be no farther than 1 in. from the edge and no higher than 2 in. above the exposure line.
- Bottom courses are doubled.
- Shakes require 18-in.-wide *interlayment* of No. 30 felt at each course.

Deck requirements

The code permits solid or spaced sheathing for wood shingles and shakes as follows:

- Spaced sheathing must be at least 1 × 4 boards spaced on center, equal to the weather exposure, to coincide with the placement of fasteners.

- Solid sheathing is required under ice barrier materials.
- Deck slope must be at least 3:12.

Underlayment
Use approved asphalt-saturated organic felt or other approved material.

Valleys
Valleys require at least No. 26 gage (0.019 in.) corrosion-resistant sheet metal extended 10 in. from the centerline each way for shingles and 11 in. from the centerline each way for shakes. End laps are at least 4 in.

Reroofing

New roof coverings shall not be installed without first removing existing roof coverings if any of the following conditions are true:
- The existing roof is water-soaked or deteriorated such that it is not adequate as a base for additional roofing.
- The existing roof covering is wood shake, slate, clay, cement, or asbestos-cement tile.
- The existing roof has two or more applications.

Note: Metal panel, metal shingle, and concrete and clay tile roof coverings may be installed over existing wood shakes.

10

Chimneys and Fireplaces

This chapter addresses the design and construction of masonry chimneys and fireplaces, including requirements for combustion air supply, clearance to combustibles, and hearth construction. Design also must comply with other structural provisions of the code, including foundation requirements. In general, the manufacturer's instructions and listing govern the installation of approved factory-built fireplaces and chimneys.

Exterior Air Supply

Both factory-built and masonry fireplaces require an exterior air supply to ensure proper fuel combustion. An alternative is mechanical ventilation as long as it is controlled so that the indoor pressure of the room or space is neutral or positive.

Factory-built fireplaces require exterior air ducts that are a listed component of the fireplace and installed according to the fireplace manufacturer's instructions. Listed ducts for masonry fireplaces must be installed according to the terms of their listing and the manufacturer's instructions.

Exterior Air Intake

The exterior air intake must be covered with a corrosion-resistant screen of ¼ in. mesh and shall not be located in the following areas:

- Garage
- Basement
- Elevation higher than the firebox

Masonry Fireplaces

The prescriptive provisions of the IRC address structural support, approved materials, dimensions, and fire safety for masonry fireplaces.

Footings

Footings must be concrete or solid masonry at least 12 in. thick, extending at least 6 in. beyond the face of the fireplace or foundation wall on all sides.

Hearths and Hearth Extensions

Hearths and hearth extensions are constructed of concrete or masonry and must meet the following requirements:

- Supported by noncombustible materials
- Reinforced to carry their own weight and all loads.
- Hearths constructed at least 4 in. thick
- Hearth extensions constructed at least:
  - 2 in. thick
  - 16 in. in front of fireplace opening
  - 8 in. beyond each side of the fireplace opening
    Note: *If the firebox opening is at least 8 in. above the hearth extension, ⅜-in.-thick brick, concrete, stone, or tile may be used. If the fireplace opening is 6 sq. ft. or larger, the hearth must extend at least 20 in. from the front and at least 12 in. beyond each side of the fireplace opening.*

Clearances to Wood and Combustibles

Clearances to wood and other combustibles must be at least 2 in. from the front and sides, and at least 4 in. from the back face. Other requirements are as follows:

- The air space between fireplace and combustibles shall not be filled (except for required noncombustible fireblocking).
- Exposed combustible trim, flooring, sheathing, and siding is permitted to abut the masonry fireplace side walls and hearth extension if these items are at least 12 in. from the inside surface of the nearest firebox lining.

- Exposed combustible mantels or trim installed directly on the masonry fireplace front must be at least 6 in. from the fireplace opening.
- If it is within 12 in. of the fireplace opening, the trim cannot project more than ⅛ in. for each inch of distance from the opening.

Masonry Chimneys

The code prescribes construction details of masonry chimneys for proper drafting, weather protection, and safety from fire.

Seismic Requirements

Four continuous vertical No. 4 reinforcing bars are required for masonry and concrete chimneys up to 40 in. wide with a single flue located in SDC D_0, D_1, or D_2. Additional requirements are as follows:

- Two additional vertical bars for each additional flue or each additional 40 in. of width
- Vertical reinforcing enclosed in ¼ in. horizontal ties every 18 in. vertically
- Two horizontal ties at each bend in vertical bars
- Anchorage to the structure at every floor, ceiling, or roof more than 6 ft. above grade

Termination

Masonry chimneys must terminate at least 3 ft. above the roof and at least 2 ft. higher than any portion of a building within 10 ft. (fig. 10.1). The code requires a chimney cap that meets the following requirements:

- Constructed or manufactured of concrete, metal or stone materials
- Sloped to shed water
- Includes a drip edge
- Has a caulked bond break around any flue liners

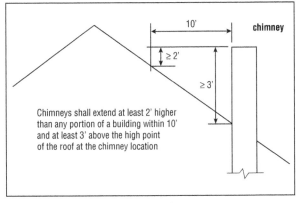

Figure 10.1 Figure height above roof

Chimney Clearances

The minimum clearances between masonry chimneys and combustible materials are as follows:

- Air space clearance to combustibles must be at least 2 in. or at least 1 in. if the chimney is located entirely outside the exterior walls of the building.
- The air space between the chimney and combustibles shall not be filled (except for required noncombustible fireblocking).
- Exposed combustible trim, and the edges of flooring, sheathing, and siding are permitted to abut the masonry chimney if these items are at least 8 in. from the inside surface of the nearest flue lining.

Chimney Fireblocking

Noncombustible fireblocking securely anchored in place is required in the concealed spaces at floors and ceilings.

Chimney Crickets

Crickets are required for chimneys 30 in. wide or larger (table 10.1, fig. 10.2).

Table 10.1 Cricket dimensions

Roof slope	Height
12:12	½ of width
8:12	⅓ of width
6:12	¼ of width
4:12	⅙ of width
3:12	⅛ of width

Factory-Built Fireplaces

Factory-built fireplaces and chimneys must be listed and labeled, tested in accordance with UL 127, and installed according to the conditions of the listing.

- Hearth extensions are to be installed in accordance with the listing of the fireplace.
- Hearth extensions must be readily distinguishable from the surrounding floor area.
- Chimney offsets cannot exceed 30 degrees from vertical.
- No more than 4 elbows may be used for chimney offsets.

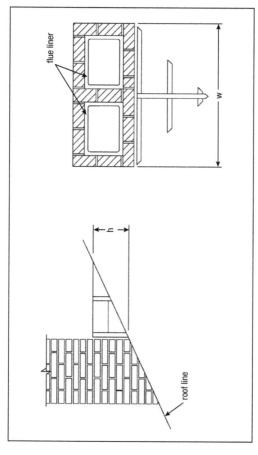

Figure 10.2 Chimney cricket

Source: International Residential Code for One- and Two-Family Dwellings, 2006. Washington, DC: International Code Council Inc., 2006, p. 310.

11

Energy Efficiency

This chapter addresses the code-prescribed methods to ensure that a home's systems and the entire building thermal envelope—including insulation, doors, windows, and cladding—conserve energy. The energy provisions of the IRC are from the *International Energy Conservation Code (IECC).*[9]

General Requirements

The IRC provides a map and lists the climate zone of each county for determining compliance with the thermal envelope requirements for dwellings. The general provisions for energy efficiency include identification and certification requirements.

Identification

Insulation materials and *fenestration* components must be labeled or identified to verify code compliance as follows:

- The factory-applied R-value mark on insulation is visible after installation.
- There are fixed markers for every 300 sq. ft. of attic space to indicate initial thickness of blown or sprayed roof/ceiling insulation.
- Windows, skylights, and glazed doors are labeled and certified by manufacturer with *U-factor* and *solar heat gain coefficient* (SHGC)

Installer Certification

For insulation products that do not bear a label or mark, the installer must provide certification listing the type, manufacturer, and R-value of insulation. The insulation installer must sign, date, and post the certificate. Certification for blown-in fiberglass or cellulose also must indicate

- Initial installed thickness
- Settled thickness
- Settled R-value
- Installed density
- Coverage area
- Number of bags installed

Permanent Certificate

The code requires the builder or registered design professional to complete a permanent energy efficiency certificate and post the certificate on a wall near the furnace or another approved location inside the building. Where posted on the electrical panel, the certificate cannot cover the service directory or other labels. The certificate must list the following:

- Predominant R-values of insulation installed at the following locations:
  - Ceiling/roof
  - Walls
  - Foundation (walls and slabs)
  - Ducts in unconditioned spaces
- Results of air-leakage testing
  - Blower door test for building envelope
  - Duct system
- U-factors and SHGC of fenestration for the following components:
  - Skylights
  - Windows
  - Doors
  - Glass block
- Type and efficiency of the following equipment:
  - Heating
  - Cooling
  - Water heating

The following appliances must be listed individually on the certificate without an efficiency designation:

- Electric furnaces
- Baseboard heaters
- Unvented gas-fired heaters

Compliance Paths

The builder may choose one of three paths for demonstrating compliance with the energy provisions as follows:

- Provisions labeled "mandatory" and those labeled "prescriptive"
- Provisions labeled "mandatory" and a simulated performance analysis using approved compliance software
- An energy rating index (ERI) approach

Building Thermal Envelope

For following the prescriptive path to compliance, the code provides the minimum thermal envelope ratings by component, based on climate zone (tables 11.1 and 11.2). In addition, the code sets mandatory requirements for reducing and measuring air leakage in the building thermal envelope.

Table 11.1 Insulation minimum R-value requirements by component

Climate zone	Ceiling R-value	Wood frame wall R-value		Floor R-value	Basement and crawl space wall R-value		Unheated slab	
		Cavity	Continuous		Continuous	Cavity	R-value	Dept (ft.)
1	30	13		13	0	0	0	
2	38	13		13	0	0	0	
3	38	20		19	5	13	0	
4 except marine	49	20		19	10	13	10	2
5 and marine 4	49	20		30	15	19	10	2
6	49	20	5	30	15	19	10	4
7 and 8	49	20	5	38	15	19	10	4

Table 11.2 Fenestration requirements by component

	Fenestration U-factor	Skylight U-factor	Glazed fenestration SHGC
1	NR	0.75	0.25
2	0.40	0.65	0.25
3	0.35	0.55	0.25
4 except marine	0.35	0.55	0.40
5 and marine 4	0.32	0.55	NR*
6	0.32	0.55	NR
7 and 8	0.32	0.55	NR

*NR: No requirement

Specific Insulation Requirements

In addition to the component requirements listed in table 11.1, the code offers alternatives to reduce the minimum R-values under specific conditions as follows:

■ A reduction in the required ceiling R-value is permitted for raised heel (energy) trusses that allow the full height of uncompressed insulation to extend over the wall top plate at the eaves.
  • R-38 is reduced to R-30.
  • R-49 is reduced to R-38.

- Ceiling insulation values greater than R-30 may be reduced to R-30 for not more than 500 sq. ft. of ceiling area if there is insufficient space for the required insulation.
- Floor insulation values greater than R-19 may be reduced to R-19 if the cavity is filled with insulation.
- The code provides insulation values for mass walls (typically above-grade concrete or concrete-block walls) and increases the required R-value when more than 50% of the insulation is on the inside of the wall.

Other specific insulation requirements are as follows:
- Slab-on-grade floors less than 12 in. below grade must be insulated inside or outside of the foundation wall.
  - Vertical and horizontal insulation may be combined to achieve the required depth of insulation.
  - Slab-edge insulation is not required in jurisdictions designated as very heavy termite infestation areas.
- Unvented crawl space wall insulation (when floor above is not insulated) must cover the wall and extend an additional 24 in. vertically or horizontally at the base.

- Exposed earth of crawl space must have a vapor retarder attached to stem walls with 6 in. sealed overlaps.
- Horizontal access hatches from conditioned to unconditioned spaces such as attics and crawl spaces require insulation, weather stripping, and a means to preserve the surrounding insulation R-value when accessing equipment.
- The maximum fenestration U-factors apply to vertical access doors to unconditioned spaces.

Air Leakage

The building thermal envelope must be sealed at the following locations:

- Air barrier joints, seams, and penetrations (including utility penetrations)
- Windows, skylights and doors
- Junctions and breaks in walls, floors and ceilings
- Garage separation
- Behind tubs and showers on exterior walls
- Access openings and doors to unconditioned spaces
- Foundation and sill plate junction
- Rim joists junctions
- Recessed lighting
- Electrical and communication boxes
- Plumbing and wiring
- HVAC register boots

- Concealed sprinklers
- Other sources of infiltration

Fireplaces
New wood-burning fireplaces require tight-fitting flue dampers or doors and outdoor combustion air. Tight-fitting doors require listing and labeling to the applicable standard.

Recessed lighting
When located in the building thermal envelope, recessed *luminaires* must meet the following criteria:
- IC (insulation contact) rated
- Labeled as meeting ASTM E 283 to reduce air leakage as follows:
  - tested at 1.57 psi pressure differential
  - no more than 2.0 *cfm* air movement
- Gasket or caulk applied between housing and interior wall or ceiling covering

Testing of the Thermal Envelope
The code requires a blower door test on every dwelling unit to verify compliance with air leakage limitations. The test pressure is 0.2 in. water gage (50 pascals). The maximum air leakage rate is
- 5 ACH in Climate Zones 1 and 2
- 3 ACH in Climate Zones 3 through 8

Systems

The IRC regulates certain components of mechanical systems and controls to conserve energy.

Thermostats

The code requires a programmable thermostat for the primary heating or cooling system.

Ducts

To conserve energy, ducts must be insulated and sealed as follows:

- Supply ducts in attics require at least R-8 insulation.
- All other ducts in unconditioned spaces require at least R-6 insulation.
- The following require sealing:
  - Ducts
  - Air handlers
  - Filter boxes

***Note:** Building framing cavities are not permitted to be supply or return air ducts.*

Duct Testing

Unless ducts and air handlers are located entirely within the building thermal envelope, the code requires pressure testing at either the rough-in stage or after construction.

Hot Water Pipe Insulation

The code requires R-3 insulation on hot water piping that

- Is located outside the conditioned space
- Travels from the water heater to a distribution manifold
- Is under a floor slab or otherwise buried
- Is part of a recirculation system other than a demand recirculation system

Mechanical Ventilation

The energy provisions require whole-house mechanical ventilation that meets the requirements of the mechanical provisions of the IRC. All ventilation fans also must be energy efficient and meet the minimum efficacy (cfm/watt) requirements of the code.

Swimming Pools

The IRC requires that heated swimming pools include these energy-conserving measures:

- A readily accessible on-off switch for pool heaters
- No continuously burning pilot lights on gas-fired heaters
- Automatic time switches for pool heaters
- A vapor retardant pool cover on the water surface

Lighting Systems

At least 75% of the lamps in permanent lighting fixtures must be *high-efficacy lamps*.

12

Mechanical Systems

This chapter addresses the mechanical (heating, ventilation, and air-conditioning, or HVAC) section of the code, which regulates the safe installation of equipment and systems that control environmental conditions of a dwelling, such as comfort heating, cooling, and ventilation, including solid- or liquid-fuel appliances. *(See chapter 13 for information on installing gas-fired appliances.)* Equipment and systems outside the scope of the IRC must comply with the applicable provisions of the *International Mechanical Code*[10] and the *International Fuel Gas Code*.[11]

General Requirements

HVAC appliances must
- Be listed and labeled for the intended application
- Carry the permanent factory-applied nameplate, including the mark of the approved testing agency

■ Include the manufacturer's operating and installation instructions attached to the appliance.

Drilling and Notching

To preserve the structural integrity of the framing system, drilling and notching of wood framing is limited to locations and dimensions as specified in chapters 5, 6, and 8 of this publication (on floors, walls, and roofs). Concealed piping that is less than 1½ in. from the nearest edge of a framing member must be protected from fastener penetration by shield plates that are at least 0.0575-in.-thick steel (No. 16 gage).

Note: Cast iron and galvanized steel pipe resist fastener penetration so they do not require additional protection.

Access to Appliances

Appliances must be accessible for service, repair, and replacement, and maintain minimum clearances as follows:

■ 30 × 30 in. working space in front of control side of the appliance
■ Furnace compartments must
  • Be at least 12 in. wider than the appliance
  • Have at least 3 in. clearance at sides, back, and top

- Maintain clearance to combustibles in accordance with their label and the manufacturer's instructions
- ▪ The access door and passageway to the appliance must be
  - At least 24 in. wide
  - Large enough to remove the largest appliance

Appliances in Attics

For maintenance or replacement of appliances installed in attics, the code requires an access opening and passageway large enough to remove the largest piece of equipment. Additional requirements are as follows:

- ▪ The attic access opening must be at least 30 × 20 in.
- ▪ The continuous solid floor of the passageway must be at least 24 in. wide.
- ▪ The passageway must be at least 22 in. wide × 30 in. high and not more than 20 ft. long.
- ▪ A receptacle outlet and switch-controlled light must be located near the appliance.
- ▪ There must be a light switch at the access opening.

Note: *An unobstructed passageway that is at least 6 ft. high may extend 50 ft.*

Appliance Installation

Installations must conform to the listing, label, and manufacturer's instructions. Installation also must include protection from ignition hazards, physical damage, and corrosion.

Appliances in Garages

For appliances installed in a garage, the IRC requires any ignition source, such as a pilot or electrical switching device, to be at least 18 in. above the floor. ***Note:*** *Appliances that are listed as flammable-vapor-ignition resistant do not require elevation of the ignition source.*

Protection from Impact

The code provides for protection of appliances against accidental physical damage from impact by vehicles. Appliances in any location subject to moving vehicles, such as in garages or outdoors adjacent to driveways, require one of the following means of protection:

- Steel pipe bollards
- Curbs
- Other approved barriers

Note: *Suspended appliances with sufficient clearance above the floor and appliances installed in an alcove out*

*of the path of vehicle travel are not subject to impact
and do not require additional barriers.*

Clearance above Grade
Appliances supported from the ground require a
level concrete slab or other approved base that is at
least 3 in. above grade.

Condensate

Condensate from cooling coils or evaporators must
drain to an approved location.

- An auxiliary drain system is required in areas
  where cooling condensate overflow would damage
  building components (such as in an attic installation above finished space). Use one of the following methods:
  - An auxiliary drain pan with a separate drain
  - A separate overflow drain line connected to the
    equipment drain pan that discharges to a con-
    spicuous location
  - Water level detection to shut off equipment,
    located either in the primary drain line or in an
    auxiliary drain pan.
- Approved condensate drain piping is at least ¾ in.
  internal diameter.

Note: *Condensate pumps in uninhabitable spaces require interconnection with the equipment to prevent appliance operation if the pump fails.*

Exhaust Systems

Mechanical exhaust systems must discharge to the outside air and are not allowed to exhaust into an attic, soffit, ridge vent, or crawl space. In addition, exhaust terminations must maintain a clearance of at least 3 ft. to openings into the building, such as windows, doors, and fresh air intakes.

Clothes Dryer Exhaust Systems

Dryer exhaust systems must
- Be independent of all other systems
- Convey moisture to the outdoors
- Terminate outside at least 3 ft. from openings into buildings or per manufacturer's instructions
- Have a backdraft damper and no screen at the exhaust termination

The dryer exhaust duct system must meet the following requirements:
- 4 in. nominal in diameter
- At least 0.016-in.-thick smooth, rigid metal
- Joints pointed in the direction of air flow

- Mechanical fastening with penetration into the duct of not more than ⅛ in.
- Sealed joints
- Duct supports spaced not more than 12 ft.

Note: *The code permits a single listed and labeled transition duct not more than 8 ft. long between the dryer and the exhaust duct. The transition duct cannot be concealed.*

Dryer Exhaust Duct Installation

The code requires the following for dryer duct installation:

- No longer than 35 ft. (unless the manufacturer permits a longer length)
- Length reductions for fittings per table 12.1
- Permanent label or tag identifying the length of duct exceeding 35 ft.
- Protection of a concealed duct that is less than 1¼ in. from the edge of framing

Note: *The code permits the installation of a listed dryer exhaust duct power ventilator (booster fan) to increase the length of the dryer duct in accordance with the manufacturer's instructions.*

Kitchen Range Hoods

Domestic open-top broiler units require an overhead metal exhaust hood unless the unit is equipped with a listed integral system. This hood is not required

Table 12.1 Dryer exhaust duct fitting equivalent length

Dryer exhaust duct fitting type	Equivalent length	
	(ft.)	(in.)
4" radius mitered 45° elbow	2	6
4" radius mitered 90° elbow	5	0
6" radius smooth 45° elbow	1	0
6" radius smooth 90° elbow	1	9
8" radius smooth 45° elbow	1	0
8" radius smooth 90° elbow	1	7
10" radius smooth 45° elbow	0	9
10" radius smooth 90° elbow	1	6

for other domestic kitchen cooking appliances. However, when it is installed, it must comply with manufacturer's instructions. Kitchen exhaust rates must comply with table 12.2. Other specific requirements follow:

- A single-wall, smooth, airtight duct of galvanized steel, stainless steel, or copper (*PVC* pipe permitted for downdraft systems) is required.
- The exhaust system cannot terminate in crawl space or attic.
- The duct must terminate outdoors with a backdraft damper.

Note: Listed ductless range hoods are not required to discharge to outdoors.

Makeup Air

For hood systems capable of exhausting more than 400 cfm, the code requires mechanical or natural *makeup air* at a rate approximately equal to the exhaust rate.

Mechanical Ventilation

When installed, local exhaust and whole-house mechanical ventilation must comply with the minimum ventilation airflow rates of the code.

Bathrooms and Toilet Rooms with Mechanical Ventilation

Local exhaust systems must meet the following requirements:

- Exhaust must terminate outdoors.
- Exhaust cannot recirculate.
- Ventilation rates must comply with table 12.2.

Table 12.2 Minimum required local exhaust rates

Area to be exhausted	Exhaust rates
Kitchens	100 cfm intermittent or 25 cfm continuous
Bathrooms and toilet rooms	Mechanical exhaust capacity of 50 cfm intermittent or 20 cfm continuous

Whole-House Mechanical Ventilation System

The energy provisions require a whole-house mechanical ventilation system for every dwelling unit. The code permits local exhaust or supply fans, or a combination, to serve as such a system. An outdoor air duct connected to the return side of an air handler is one approved method for providing supply ventilation. The whole-house ventilation system must supply air at the minimum continuous rate specified (table 12.3) or the minimum intermittent rate (table 12.4).

Table 12.3 Continuous whole-house mechanical ventilation system airflow rate requirements

Dwelling unit floor area (sq. ft.)	Number of bedrooms				
	0–1	2–3	4–5	6–7	> 7
	Airflow in CFM				
< 1,500	30	45	60	75	90
1,501–3,000	45	60	75	90	105
3,001–4,500	60	75	90	105	120
4,501–6,000	75	90	105	120	135
6,001–7,500	90	105	120	135	150
> 7,500	105	120	135	150	165

Table 12.4 Intermittent whole-house mechanical ventilation rate factors

Run-time percentage in each 4-hour segment	25%	33%	50%	66%	75%	100%
Factor	4	3	2	1.5	1.3	1.0

Note: The code permits intermittent operation when the system can be set to operate at least 25% of each 4-hour segment and the ventilation rate prescribed in table 12.3 is multiplied by the factor in table 12.4.

Duct Systems

The IRC sets requirements for the materials, fabrication, and installation of duct systems as follows:

- Approved materials meeting the temperature and flame-spread ratings of the code
- Factory-made ducts listed and labeled to applicable UL standards
- Airtight joints and seams
- Maximum flame-spread of 25 and smoke-developed index of 50 for duct insulation
- Labels for external duct insulation with R-value, flame-spread rating, and smoke-developed index
- Maximum 10 ft. spacing of metal duct supports
  - ½ in. wide, 18-gage metal straps

    or
  - 12-gage galvanized wire

Note: The code also permits the use of spray polyurethane foam for insulating and sealing ducts when the foam meets the applicable conditions and specifications.

Return Air Openings

The code restricts the sources of return air to prevent the circulation of unpleasant or noxious odors and to prevent negative pressures in confined spaces or appliance locations. Return air shall not be obtained from the following locations:

- Less than 10 ft. from an open combustion chamber or draft hood of another appliance in the same space
- Another dwelling unit
- Closet, bathroom, *toilet room*, kitchen, garage, mechanical room, boiler room, furnace room or unconditioned attic
- Unconditioned crawl space through a direct connection

Note: *The code does not prohibit transfer openings in the crawl space enclosure.*

The amount of return air taken from any room shall not exceed the flow rate of supply air to that room.

Combustion Air

Fuel-burning appliances require a supply of air for fuel combustion, draft hood dilution, and ventilation of the space in which the appliance is installed. Combustion air requirements for solid-fuel-burning appliances must follow the manufacturer's installation instructions. Oil-fired appliances require combustion air in accordance with National Fire Protection Association (NFPA) 31.[12] The requirements for combustion and dilution air for gas-fired appliances are in chapter 13 of this publication.

Chimneys and Vents for Solid-Fuel and Oil-Burning Appliances

The code prohibits connecting a solid-fuel-burning appliance or fireplace to a chimney that vents another appliance.

Ranges and Ovens

The IRC regulates the installation of household cooking appliances as follows:

- Freestanding or built-in ranges require at least 30 in. of clearance from unprotected combustible material located above the cooking top.
- A reduced clearance is permitted if it complies with the listing and labeling of the range hood or appliance.
- Cooking appliances must be listed and labeled for household use, and installed according to the manufacturer's instructions.

13

Fuel Gas

This chapter addresses the safe installation of gas-fired appliances and fuel-gas piping systems, including requirements for combustion and ventilation air, venting, and connections. (Liquid- and solid-fuel appliances are covered in chapter 12.) The IRC includes excerpts from the *International Fuel Gas Code*,[13] which may be used for installations outside the scope of the IRC.

General

Appliance installation requirements include provisions for preventing damage to building components.

- Appliances must be listed and labeled for the specific intended purpose.
- An auxiliary drain pan is required for Category IV condensing appliances where overflow would damage any part of the building, such as where a furnace is installed in an attic above living space.

Note: *The auxiliary pan is not required if the appliance has a device to shut the appliance down automatically if the primary drain becomes clogged.*

Note: *Condensate pumps in uninhabitable spaces such as attics or crawl spaces require interconnection with the equipment to prevent appliance operation if the pump fails.*

Appliance Location

In addition to the code requirements, appliance location depends on the conditions of its listing. Gas-fired appliances installed in sleeping rooms, bathrooms, toilet rooms, or storage closets, or in a space that opens only into these rooms or spaces, are limited to the following specific installations. Otherwise, gas-fired appliances are prohibited in these rooms or areas.

- *Direct-vent appliances* are permitted if they are installed according to the manufacturer's instructions.
- Vented room heaters, wall furnaces, vented decorative appliances, and vented gas fireplaces are permitted in rooms that meet the required volume for indoor combustion air.
- A single unvented room heater may be permitted in a bathroom or bedroom, subject to input

rating limitations, combustion air volumes, and required automatic oxygen depletion safety shut-off systems.

▨ An appliance is permitted in a room or space that opens only into a bedroom or bathroom under the following conditions:
  • The space is used for no other purpose.
  • A solid, weather-stripped, self-closing door is installed.
  • All combustion air is taken directly from the outdoors.

Combustion, Ventilation, and Dilution Air

Gas-fired appliances require an air supply for fuel combustion, draft hood dilution, and ventilation of the space in which the appliance is installed. Combustion air requirements for solid- and liquid-fuel-burning appliances appear in the mechanical section of the code and are covered in chapter 12 of this publication.

▨ In addition to the code requirements, combustion, ventilation, and dilution air is required in accordance with the appliance manufacturer's instructions.

▨ Exhaust fans, ventilation systems, clothes dryers, and fireplaces require makeup air.

Indoor Combustion Air

The code permits combustion air to be obtained from inside the building under the following conditions:

- The minimum required volume of air is 50 cu. ft. per 1,000 *Btu/h*.
- Openings used to connect indoor spaces to obtain the required volume of indoor combustion air must be sized and located as follows (fig. 13.1):
  - Each opening requires a free area of 1 sq. in. per 1,000 Btu/h and at least 100 sq. in.
  - One opening must begin within 12 in. of the top and one opening must begin within 12 in. of the bottom of the enclosure.

Outdoor Combustion Air Obtained through Two Openings or Ducts

This option for necessary combustion air connects the space containing the appliance to the outdoors through two openings or ducts.

- One opening is to begin within 12 in. of the top of the enclosure and one within 12 in. of the bottom (fig. 13.2).
- Openings are permitted to connect to spaces directly communicating with the outdoors, such as ventilated crawl spaces or ventilated attic spaces.

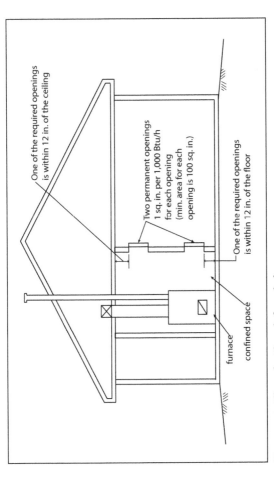

Figure 13.1 All combustion air from indoors

Source: 2006 International Residential Code Study Companion. Washington, DC: International Code Council, 2006, p.290.

- ▪ Sizes of openings are as follows:
  - For direct openings to outdoors or vertical ducts, the free area of each duct must be at least 1 sq. in. per 4,000 Btu/h of total input rating.
  - For horizontal ducts, the free area of each opening must be at least 1 sq. in. per 2,000 Btu/h of total input rating.

Outdoor Combustion Air Obtained through Single Opening or Duct

As an alternative to two openings, the code permits a single opening to supply combustion air from the outdoors, provided the opening is increased in size and meets other requirements as follows:

- ▪ The opening must begin within 12 in. of the top of the enclosure (fig. 13.3).
- ▪ Free area is to be at least 1 sq. in. per 3,000 Btu/h of the total input rating.
- ▪ Free area must equal at least the sum of the areas of all vent connectors in the space.

Combination Indoor and Outdoor Combustion Air

The calculated size of outdoor openings may be reduced by a factor based on a ratio of available interior air volume divided by the required volume of combustion air.

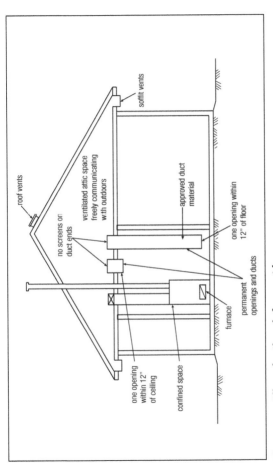

Figure 13.2 All combustion air from outdoors

Source: 2006 International Residential Code Study Companion. Washington, DC: International Code Council, 2006, p. 292.

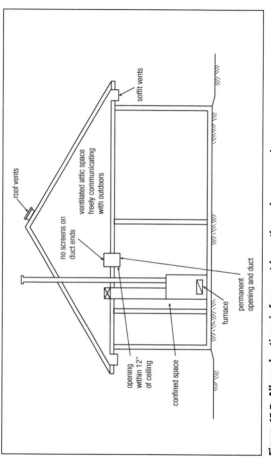

Figure 13.3 All combustion air from outdoors through one opening

Source: 2006 International Residential Code Study Companion. Washington, DC: International Code Council, 2006, p. 292.

Mechanical Combustion Air Supply

As an alternative to natural infiltration of combustion air, the code permits a mechanical system to bring in combustion air when the following conditions are met:

- Combustion air must be supplied from the outdoors at a rate of at least 0.35 cfm per 1,000 Btu/h.
- Exhaust fans require makeup air.
- Appliances must interlock with the mechanical air supply system.

Louvers and Grilles Net Free Area

When using louvers or grilles to satisfy combustion air requirements, the amount of net free area is as specified by the manufacturer. If the manufacturer's specifications are not known, the amount can be calculated as follows:

- 75% of the gross area for metal louvers
- 25% of the gross area for wood louvers

Combustion Air Ducts

Combustion air ducts must

- Be manufactured of galvanized steel
- Serve only lower or only upper openings (not both)
- Have at least 12 in. clearance above grade for exterior air intake openings

Appliances in Garages

The following requirements apply to appliances installed in garages:

- The appliance must be protected from automobile impact by a barrier, elevation, placement in an offset alcove, or by other approved means.
- The ignition source for the appliance must be at least 18 in. above the floor.

Note: The code does not require elevation for appliances that are listed as flammable-vapor-ignition resistant.

Exterior Equipment and Appliance Installation

Equipment and appliances installed outdoors must be

- Supported on a level concrete slab or other approved material extending at least 3 in. above grade, or
- Suspended at least 6 in. above the finished grade

Fuel-Gas Piping

The IRC provides for the design, materials, and safe installation of fuel-gas piping to serve gas-fired appliances.

Pipe Sizing

Gas piping is sized to supply adequate volume to meet the demand of the connected appliances. Appliance input ratings, type of gas (natural or liquid petroleum), type of pipe, length of pipe, inlet pressure, and pressure drop are among the factors that determine proper pipe size. The code includes pipe sizing tables and sizing equations, but permits other approved methods for sizing gas pipe.

Piping Materials

The IRC permits any of the following materials for gas piping:

- Schedule 40 steel
- Approved seamless metallic tubing if gas used is not corrosive to the material
- Corrugated stainless steel tubing (CSST)
- Approved plastic pipe, tubing, and fittings for outdoor underground installations

Piping System Prohibited Locations

The code does not permit gas piping in the following locations:

- Supply, return, or exhaust duct
- Clothes chute
- Chimney or gas vent

- Through any townhome unit other than the unit being served

Shield Plate Requirements

Concealed piping installed less than 1½ in. from the nearest edge of the framing member must be protected from fastener penetration by shield plates that are at least 00.0575-in.-thick (No. 16 gage) steel. ***Note:*** *Black and galvanized steel pipe do not require protection.*

Other Installation Requirements

Gas piping must be inspected and pressure tested before it is concealed or put into service. The following requirements apply to gas piping in outdoor or underground locations:

- Aboveground piping outdoors must be at least 3½ in. above ground and above roof surfaces.
- Exposed exterior ferrous metal piping must be protected from corrosion.
- Underground piping must be approved for the location or wrapped with approved material. ***Note:*** *Galvanizing is not considered adequate protection.*
- Underground piping must be buried at least 12 in.
- Gas piping must not penetrate foundation walls underground.

Appliance Connections

Connectors are not allowed to pass through walls, floors, partitions, ceilings, or appliance housings (other than connectors to fireplace inserts with proper grommets in accordance with the manufacturer's instructions). Connector length is limited to no more than 6 ft. The following are approved for appliance connection to the gas piping system:

- Rigid metallic pipe and fittings
- CSST
- Listed and labeled appliance connectors

Shutoff Valve

Each appliance requires a shutoff valve and, except for alternate provisions that apply to decorative appliances and fireplaces, the valve is to be

- Upstream of the connector
- Accessible and in the same room as the appliance
- Within 6 ft. of the appliance

Note: In a manifold system, the shutoff valve may be located at the manifold and up to 50 ft. from the appliance when the shutoff valve is readily accessible and permanently identified.

Sediment Trap

A sediment trap is required downstream of the shutoff valve and adjacent to the inlet of equipment, except for

- Illuminating appliances
- Ranges
- Clothes dryers
- Outdoor grills
- Gas fireplaces

Piping Support

See table 13.1 for maximum spacing of supports for gas piping. CSST supports must follow the manufacturer's instructions.

Vents

Gas appliances must vent to the atmosphere to release their combustion products. Vents must

Table 13.1 Support of fuel-gas piping

Steel pipe, nominal size of pipe (in.)	Spacing of supports (ft.)	Nominal size of tubing smooth-wall (in., outside diameter)	Spacing of supports (ft.)
½	6	½	4
¾ or 1	8	⅝ or ¾	6
1¼ or larger (horizontal)	10	⅞ or 1 (horizontal)	8
1¼ or larger (vertical)	Every floor level	1 or larger (vertical)	Every floor level

- Be listed and labeled for use with the type of appliance (except plastic piping used to vent *Category IV appliances*)
- Have appropriate clearance to combustibles as specified by the manufacturer
- Have a 0.0187 in. (26 gage) sheet metal insulation shield that terminates at least 2 in. above insulation in attic if vents pass through insulated assemblies

Where vents are installed in concealed locations through holes or notches in framing, and are less than 1½ in. from the edge of the framing, the vents require protection from fastener penetration by installing a fastener shield plate of 00.057-in.-thick (16 gage) steel. The shield plate must

- Cover the vent area at the framing
- Ensure at least 4 in. of protection above the bottom plates, below the top plates, and to each side of a stud, joist, or rafter

Gas Vent Roof Termination

The required termination height for gas vents of no more than 12 in. and at least 8 ft. from a vertical wall is based on the roof pitch (table 13.2 and fig. 13.4). Gas vents larger than 12 in. or less than 8 ft. from a vertical wall must terminate at least 2 ft.

**Table 13.2 Gas vent termination locations
(Minimum height from roof to lowest discharge opening) for
listed caps 12 in. or smaller, at least 8 ft. from a vertical wall**

Roof slope	Minimum height in ft.
Flat to 6/12	1.0
> 6/12 to 7/12	1.25
> 7/12 to 8/12	1.5
> 8/12 to 9/12	2.0
> 9/12 to 10/12	2.5
> 10/12 to 11/12	3.25
> 11/12 to 12/12	4.0
> 12/12 to 14/12	5.0
> 14/12 to 16/12	6.0
> 16/12 to 18/12	7.0
> 18/12 to 20/12	7.5
> 20/12 to 21/12	8.0

above the roof and any portion of a building that is
within 10 ft. horizontally. In addition, gas vent roof
termination must include

- Roof flashing or roof jack
- Listed termination cap

Note: Vent terminals require a minimum 12 in. horizontal clearance from the swing of a door.

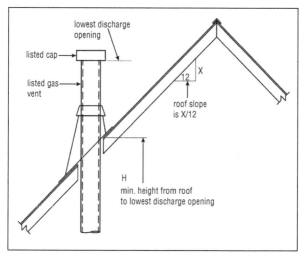

Figure 13.4 Gas vent termination locations (for listed caps 12 in. or smaller, at least 8 ft. from a vertical wall)

Source: International Residential Code for One- and Two-Family Dwellings, 2006. Washington, DC: International Code Council Inc., 2006, p. 400.

Mechanical Draft Venting System Termination

Mechanical draft venting systems must terminate at least 3 ft. above any forced-air inlet located within 10 ft. For other than direct vent appliances, mechanical draft venting systems must meet the following termination clearances:

- At least 4 ft. below, 4 ft. horizontally from, and 1 ft. above any door, operable window, or gravity air inlet in the building
- At least 12 in. above grade

Direct-Vent Appliance Vent Termination

Direct-vent appliance vents must terminate at least

- 6 in. from any air opening into the building when appliance input is less than or equal to 10,000 Btu/h
- 9 in. from any air opening into the building when appliance input is greater than 10,000 but not more than 50,000 Btu/h
- 12 in. from any air opening into the building when appliance input is more than 50,000 Btu/h

Gas Clothes Dryer Exhaust

The system must follow manufacturer's instructions, be independent of all other systems, and

- Convey moisture and combustion products to the outside
- Have an outside termination equipped with a backdraft damper and no screens
- Include the required makeup air for exhaust greater than 200 cfm or for closet enclosures

- Have a nominal 4 in. diameter smooth, rigid metal dryer duct of at least 0.016-in.-thick material supported at 4 ft. intervals
- Have duct ends inserted into adjoining duct or fitting in the direction of air flow
- Not have screws or fasteners penetrating into duct more than 1/8 in.

Note: *The code permits a single listed and labeled transition duct no longer than 8 ft. to connect the dryer to the exhaust duct system. The transition duct cannot be concealed within construction.*

Protection of Dryer Exhaust Duct

The code requires shield plates to protect dryer exhaust duct from nail or screw penetration where there is less than 1¼ in. between the duct and the face of the framing member. Shield plates must comply with the following:

- Be constructed of steel at least 0.062 in. thick
- Extend at least 2 in. above sole plates and below top plates

Dryer Exhaust Duct Length

Unless the manufacturer's instructions permit a longer length, the IRC limits a gas dryer exhaust duct to no more than 35 ft. This length is reduced for each fitting based on the size and type of fitting *(refer to table 12.1 in previous chapter)*. For a dryer exhaust duct that exceeds 35 ft., the installed length must be identified on a permanent tag within 6 ft. of the dryer.

Note: *The code permits the installation of a listed dryer exhaust duct power ventilator (booster fan) to increase the length of the dryer duct in accordance with the manufacturer's instructions.*

Forced-Air Furnace Return Air/ Outside Air Prohibited Sources

For a forced-air furnace, the code prohibits the following locations as sources of return or outside air:

- Less than 10 ft. from an appliance vent outlet, a plumbing vent, or an exhaust fan outlet, unless the outlet is 3 ft. above the outside air inlet
- Less than 10 ft. above a public way or driveway
- Rooms without an adequate volume of air
- A closet, bathroom, toilet room, kitchen, garage, mechanical room, furnace room, attic, or another dwelling unit
- A crawl space by means of direct connection to the return side of a forced air system
- A room or space containing a non-direct-vent, fuel-burning appliance, unless specific volume

and location requirements are met, if that room or space is the sole source of return air

Suspended Unit Heaters

Unless the listing and manufacturer's instructions permit reduced clearances, suspended-type unit heaters must have clearances to combustible materials of at least

- 18 in. from the sides
- 12 in. from the bottom
- 6 in. from the top

Cooking Appliances

The code requires cooking appliances installed in dwelling units to be listed and labeled as household-type appliances for domestic use. Commercial appliances are not permitted unless they are also listed as domestic cooking appliances. The minimum vertical clearance between the cooking surface and combustible materials or metal cabinets above is 30 in. The code permits a clearance of 24 in. for the following:

- Underside of combustible material or metal cabinet covered with ¼ in. insulating millboard covered with sheet metal

- Metal ventilating hood with a ¼ in. space between hood and combustible material or metal cabinet
- Microwave or other cooking appliance listed for 24 in. clearance above cooking surface

14

Plumbing

This chapter covers plumbing system design and installations typical of dwellings. Methods and materials outside the scope of the IRC must comply with the *International Plumbing Code* (IPC).[14]

Water and Sewer Requirements

The water distribution and drainage system of any building must meet the following minimum requirements and be connected to

- An approved public water supply, if available, or individual private water supply such as a well, if the public water supply is not available
- An approved public sewage disposal system or individual private system, such as an on-site septic system, if a public system is not available

General Plumbing Requirements

To perform as intended and to prevent damage or contamination, piping requires adequate support and protection (table 14.1). Concealed piping installed through holes or notches in studs, joists, rafters, or similar members and less than 1¼ in. from the nearest edge of the member requires protection from fastener penetration by shield plates (fig. 14.1).

- Protective shield plates must be at least 0.057-in.-thick (No. 16 gage) steel.
- Plates must cover the area where the pipe passes through the member.
- Shield plates must extend at least 2 in. above sole plates and below the top plates of wall framing.

Note: Cast-iron and galvanized steel pipe do not require protection.

Other general requirements for protection of plumbing piping follow:

- Pipes passing through foundation walls require a pipe sleeve.
- Water, drain, and sewer pipe must be protected from freezing
- Water service pipe is buried at least 12 in. deep and at least 6 in. below the frost line.
- In-ground piping requires continuous support on suitable bedding materials.

Table 14.1 Piping support

Piping material	Maximum horizontal spacing (ft.)	Maximum vertical spacing (ft.)
ABS pipe	4	10
Cast-iron pipe, < 10 ft. lengths of pipe	5	15
Cast-iron pipe, 10 ft. lengths of pipe	10	15
Copper or copper alloy pipe	12	10
Copper or copper alloy tubing (1¼ in. and smaller)	6	10
Copper or copper alloy tubing (1½ in. and larger)	10	10
Cross-linked polyethylene (PEX) pipe (1 in. and smaller)	2.67	10
Cross-linked polyethylene (PEX) pipe (1¼ in. and larger)	4	10
Cross-linked polyethylene/aluminum/cross-linked polyethylene (PEX-AL-PEX) pipe	2.67	4
CPVC pipe or tubing (1 in. and smaller)	3	10
CPVC pipe or tubing (1¼ in. and larger)	4	10
Polybutylene (PB) pipe or tubing	2.67	4
Polyethylene of raised temperature (PE-RT) pipe (1 in. and smaller)	2.67	10
Polyethylene of raised temperature (PE-RT) pipe (1¼ in. and larger)	4	10
Polypropylene (PP) pipe or tubing (1 in. and smaller)	2.67	10
Polypropylene (PP) pipe or tubing (1¼ in. and larger)	4	10
PVC pipe	4	10
Stainless steel drainage systems	10	10
Steel pipe	12	15

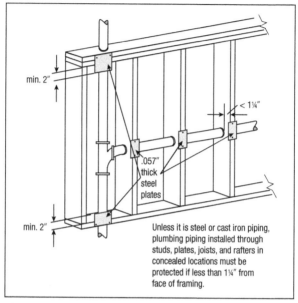

min. 2″

< 1¼″

.057″ thick steel plates

min. 2″

Unless it is steel or cast iron piping, plumbing piping installed through studs, plates, joists, and rafters in concealed locations must be protected if less than 1¼″ from face of framing.

Figure 14.1 Protection of piping against physical damage

■ Backfill over pipe must be free of debris, rocks, concrete, and frozen material.

Water Supply System Design Criteria

The system must be designed and installed to deliver adequate water volume and pressure for plumbing fixtures to operate efficiently and properly.

- Approved static pressure for the water service at the building entrance is 40–80 psi.
- Water service pipe must be at least ¾ in.
- Water supply fixture unit values, developed length of piping, and water pressure determine pipe size.

Valves

The IRC prescribes the locations and types of valves required for water distribution systems.

- An accessible main shutoff valve must be located near the entrance of the water service.
- A readily accessible full-open valve is required at each water heater on the cold-water supply pipe.
- An accessible individual shutoff valve is required on the fixture supply pipe to each plumbing fixture other than bathtubs and showers.
- Hose bibbs subject to freezing require an accessible stop-and-waste-type valve inside the building. This valve is not required for frost-proof hose bibbs that extend into heated space.

Water Service Installation

The water service must be protected from potential contamination when installed near the *building sewer*.

- Water service pipe is permitted in the same trench with a building sewer if the sewer pipe is listed for underground use within a building.

- For other types of building sewer pipe, the water service must have at least a 5 ft. horizontal separation or be installed on a ledge at least 12 in. above and to one side of the highest point of the sewer.

Protection of Potable Water Supply from Contamination

To protect the potable water supply, a listed backflow prevention device is required for the following:

- Hose connections
- Boilers and heat exchangers
- Lawn irrigation systems

An *air gap* is another means to prevent backflow. It is used for the water outlets of fixtures (fig. 14.2). The minimum air gap varies according to fixture type and application (table 14.2).

Dwelling Unit Fire Sprinkler System

The IRC requires an automatic fire sprinkler system in all new one- and two-family dwellings and townhomes. In the plumbing provisions, the code provides a simple, prescriptive approach to the design of dwelling fire sprinkler systems that is an approved alternative to NFPA 13D. The system may be multi-purpose—supplying both sprinklers and plumbing fixtures—or a stand-alone system

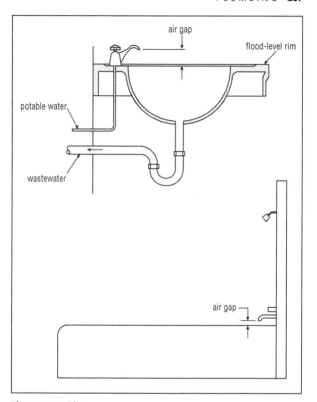

Figure 14.2 Air gap

Source: 2006 International Residential Code Study Companion. Washington, DC: International Code Council, 2006, p. 367.

Table 14.2 Minimum air gaps

Fixture	Minimum air gap	
	Away from a wall (in.)	Close to a wall (in.)
Effective openings greater than 1 in.	Two times the diameter of the effective opening	Three times the diameter of the effective opening
Lavatories and other fixtures with effective opening not greater than ½" in diameter	1	1.5
Over-rim bath fillers and other fixtures with effective openings not greater than 1" in diameter	2	3
Sink, laundry trays, gooseneck back faucets, and other fixtures with effective openings not greater than ¾" in diameter	1.5	2.5

independent of the water distribution system. In either case, a dwelling fire sprinkler system does not require a fire department connection. The code permits any combination of water supply systems to meet the required dwelling fire sprinkler system capacity including the following:

- Domestic water supply
- Well

▪ Elevated storage tank
▪ Approved pressure tank
▪ Stored water source with an automatic pump

Automatic sprinkler protection is not required throughout a one- and two-family dwelling or townhome. Generally, sprinklers are not required in the following areas:

▪ Closets with areas of 24 sq. ft. or less and not more than 3 ft. deep
▪ Bathrooms with areas of 55 sq. ft. or less
▪ Exterior porches
▪ Garages and carports
▪ Attics
▪ Crawl spaces
▪ Concealed spaces not intended for or used as living spaces

Note: *In attics, crawl spaces, and other concealed unoccupied spaces that contain fuel-fired equipment, the code requires a single sprinkler above the equipment.*

Non-potable Water Systems

Provisions for collecting, storing, and using various types of non-potable water recognize the growing need for water conservation and the increased implementation of water conservation programs in many regions of the United States. The code addresses the following non-potable water systems:

- Rainwater
- Reclaimed water
- *On-site non-potable water reuse systems*

Note: *On-site non-potable water reuse systems include but are not limited to gray water systems. Gray water is the waste from lavatories, bathtubs, showers, washing machines and laundry tubs.*

Non-potable water is not safe for drinking, bathing, or cooking purposes, and the code requires clear warning signs at non-potable water outlets such as hose connections and yard hydrants. Other general requirements follow:

- Filtration
- Disinfection as necessary
- Approved storage tanks
- Valves at prescribed locations
- Purple distribution piping
- Piping and storage tank labeling

Non-potable water is typically used for flushing water closets and urinals and for landscape irrigation. The code does not require disinfection of non-potable water used for subsurface landscape irrigation.

Sanitary Drainage

Proper operation of the sanitary drainage system depends on adequate pipe size of approved materials, appropriate slope and support, transition fittings suitable for the location (table 14.3), adequate cleanouts, and proper venting. Total *d.f.u.* of the fixtures served determine pipe sizes (tables 14.4–14.6).

Table 14.3 Fittings for change in direction

Type of fitting pattern	Change in direction		
	Horizontal to vertical	**Vertical to horizontal**	**Horizontal to horizontal**
Sixteenth bend	X	X	X
Eighth bend	X	X	X
Sixth bend	X	X	X
Quarter bend	X	X*	X*
Short sweep	X	X*†	X*
Long sweep	X	X	X
Sanitary tee	X	—	—
Wye	X	X	X
Combination wye and eighth bend	X	X	X

*The fittings shall be permitted only for a 2 in. or smaller fixture drain
†3 in. and larger

Table 14.4 Drainage fixture unit (d.f.u.) values for various plumbing fixtures

Type of fixture or group of fixtures	Drainage fixture unit value (d.f.u.)
Bar sink	1
Bathtub (with or without shower head and/or whirlpool attachments)	2
Bidet	1
Clothes washer standpipe	2
Dishwasher	2
Floor drain	0
Kitchen sink	2
Lavatory	1
Laundry tub	2
Shower stall	2
Water closet (1.6 gal. per flush)	3
Water closet (greater than 1.6 gal. per flush)	4
Full-bath group with bathtub (with 1.6 gal. per flush water closet and with or without shower head and/or whirlpool attachment on the bathtub or shower stall)	5
Full-bath group with bathtub (water closet greater than 1.6 gal. per flush, and with or without shower head and/or whirlpool attachment on the bathtub or shower stall)	6
Half-bath group (1.6 gal. per flush water closet plus lavatory)	4
Half-bath group (water closet greater than 1.6 gal. per flush plus lavatory)	5

(continued)

Table 14.4 Drainage fixture unit (d.f.u.) values for various plumbing fixtures (*continued*)

Type of fixture or group of fixtures	Drainage fixture unit value (d.f.u.)
Kitchen group (dishwasher and sink with or without food-waste disposer)	2
Laundry group (clothes washer standpipe and laundry tub)	3
Multiple-bath groups:	
1.5 baths	7
2 baths	8
2.5 baths	9
3 baths	10
3.5 baths	11

Cleanouts

Cleanouts are required in horizontal drain lines at each change of direction greater than 45°. Where more than one change of direction occurs, only one cleanout is required in each 40 ft. A readily removable fixture or trap may serve as a cleanout. Other cleanout requirements follow:

- Cleanouts must have accessible clearance of at least 18 in. for pipes of 3 in. or more, and 12 in. for smaller pipes.
- Cleanouts must allow cleaning in the direction of the flow.

Table 14.5 Maximum fixture units allowed to be connected to branches and stacks

Nominal pipe size (in.)	Any horizontal fixture branch	Any one vertical stack or drain
1¼	—	—
1½	3	4
2	6	10
2½	12	20
3	20	48
4	160	240

Note: 1¼ in. pipe is only permitted for a single-fixture drain or trap arm (lavatory). Water closets are not permitted on drain lines less than 3 in.

Table 14.6 Maximum number of fixture units allowed to be connected to building drain, building drain branches, or building sewer

Diameter of pipe (in.)	Slope per ft.		
	⅛ in.	¼ in.	½ in.
2	—	21	27
2½	—	24	31
3	36	42	50
4	180	216	250

Note: Water closets are not permitted on drain lines smaller than 3 in.

- The required cleanout at the junction of the *building drain* and building sewer may be inside or outside the building. A two-way cleanout is permitted outside the building to serve both the sewer and the building drain.

Minimum Slope for Horizontal Drainage Piping
The code sets minimum slope requirements for drainage piping based on the pipe size as follows:
- ¼:12 slope for pipe sizes 2½ in. or less
- ⅛:12 slope for pipe sizes 3 in. or more

Vents

Venting provides air to equalize pressure for proper operation of the drainage system, prevents siphoning of water seals, and allows the safe escape of any sewer gas to the outside atmosphere.
- Every trap and trapped fixture must be vented to protect the trap seal.
- At least one vent pipe to the outdoors is required. It must be sized appropriately for the building drain size.

Vent Termination
Proper plumbing vent termination ensures proper operation of the vent, maintains weather resistance

of the building, and prevents sewer gas from entering the building.

- Vent pipe termination is at least 6 in. above the roof or 6 in. above the anticipated snow accumulation.
- Vent pipe termination is at least 7 ft. above a roof used for an observation deck or sunbathing deck or for similar purposes.
- Vent extensions through the roof in cold climates that are based on regional outside design temperature require at least 3-in.-diameter pipe, starting 1 ft. inside the building thermal envelope, for protection against frost closure.
- Vents through the roof require approved flashings that are weather tight.
- Minimum clearances from vent terminations to any door, openable window, or other air intake opening are 4 ft. below, 10 ft. horizontally from, or 3 ft. above the opening.

Vent Connections and Grades

The following requirements prevent stoppages and trapping of water in the vent system:

- Vents require adequate slope to allow moisture and condensate to drain back to soil and waste pipes.

- Dry vent connection to a horizontal drain must be above the centerline of the drain.
- Dry vents must to rise vertically to a point at least 6 in. above the flood rim of the fixture.
- Branch vent connection to a vent stack must be at least 6 in. above the flood level of the fixture served.
- If plumbing is roughed in for future fixtures, a vent must be installed to serve those fixtures.

Fixture Vents

Proper slope and maximum lengths for fixture drains ensure proper exchange of air and prevent siphoning of the fixture trap.

- Total fall in a fixture drain to the vent connection cannot exceed one pipe diameter.
- Vent connection to fixture drain is not permitted to be below the trap weir.
- Distance from the trap to the vent is limited as shown in table 14.7, except for self-siphoning fixtures such as water closets (fig. 14.3).

Common Vent

A common vent is a single pipe venting two trap arms of fixtures on the same floor, either back to back, or one above the other (table 14.8).

Table 14.7 Maximum distance of fixture trap from vent

Size of trap (in.)	Slope (in. per ft.)	Distance from trap (ft.)
1¼	¼	5
1½	¼	6
2	¼	8
3	¼	12
4	¼	16

- Where the drains connect at the same level, the vent connection shall be at the interconnection of the fixture drains or downstream from the interconnection.
- Where the fixture drains connect at different levels, the upper fixture cannot be a water closet. Vent connection is not permitted upstream from the horizontal drain connections.

Wet Venting

Vents are required to equalize the pressure in the *DWV* system to preserve the water seal at traps, and to provide air for proper drainage. Wet venting is based on the principle that drain pipe for a fixture or fixtures will have adequate amounts of air to also serve as the vent for a downstream fixture. The

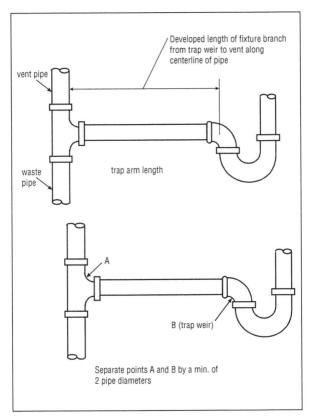

Figure 14.3 Trap arm length to vent

Source: 2006 International Residential Code Study Companion. Washington, DC: International Code Council, 2006, p. 389.

Table 14.8 Common vent sizes

Pipe size (in.)	Maximum discharge from upper fixture drain (d.f.u.)
1½	1
2	4
2½ to 3	6

section of pipe serving as a wet vent (acting as both a drain and a vent) must be sized adequately and is limited to a lower d.f.u. load than would be allowed for drainage piping (table 14.9).

- Horizontal wet venting is permitted for any combination of fixtures in two bathroom groups.
- Only one fixture is permitted upstream of the dry vent connection to the horizontal wet vent.

Table 14.9 Wet vent size

Wet vent pipe size (in.)	Fixture unit load (d.f.u.)
1½	1
2	4
2½	6
3	12
4	32

- The dry vent connection to the wet vent must be an individual vent or common vent to any bathroom group fixture except an emergency floor drain.

Waste Stack Vent

A *waste stack* is permitted as a vent for all connected fixtures, provided

- The waste stack is vertical without offsets between the highest and lowest fixture drain connection.
- Every fixture drain has a separate connection to the waste stack.
- The stack does not receive the discharge of water closets or urinals.
- The *stack vent* is the same size or larger than the waste stack (table 14.10).

Island Fixture Venting

Island venting is an alternative for venting isolated fixtures where other means are not feasible. Island venting must comply with the following:

- Kitchen sinks and lavatories may have island venting (fig. 14.4).
- The fixture vent loops above the level of the sink drain and back down below the floor level before continuing horizontally to a vent.

Table 14.10 Waste stack vent size

	Maximum number of fixture units (d.f.u.)	
Stack size (in.)	**Total discharge into one branch interval**	**Total discharge for stack**
1½	1	2
2	2	4
2½	No limit	8
3	No limit	24
4	No limit	50

- The horizontal section is installed as required for drainage piping and is in addition to the fixture drain.
- Cleanouts are required.

Air Admittance Valves

An air admittance valve is a one-way valve designed to allow air into the plumbing drainage system when a negative pressure develops in the piping (when there is flow from the fixture). The device is at a fixture vent terminal that pulls air from the room in which the fixture is located and is an alternative to venting to the outside air. The valve is designed to close by gravity and seal the terminal under no-flow

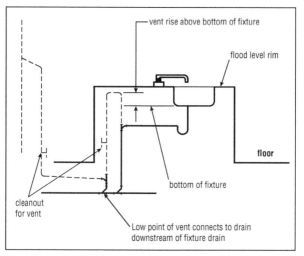

Figure 14.4 Island venting

Source: 2006 International Residential Code Study Companion. Washington, DC: International Code Council, 2006, p. 396.

conditions to prevent introducing sewer gas into the dwelling.

- Approved air admittance valves are permitted for individual, branch, circuit, and stack vents.
- Valves must be installed according to the manufacturer's instructions.
- Installation must be at least 4 in. above branch or fixture drain (fig. 14.5).

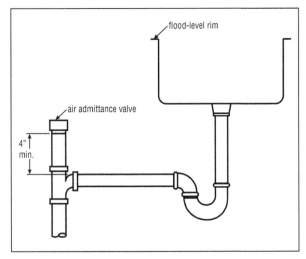

Figure 14.5 Air admittance valve

Source: 2006 International Residential Code Study Companion. Washington, DC: International Code Council, 2006, p. 397.

- Stack-type air admittance valves must be installed at least 6 in. above the flood-level rim of the highest fixture.
- Air admittance valves installed in attics must be at least 6 in. above insulation.
- The space containing the air admittance valve requires access and ventilation.
- At least one vent is required to extend outdoors to the open air.

Fixture Traps

Traps provide a water seal to prevent sewer gases from entering the building. Trap sizes vary by fixture (table 14.11). Requirements for fixture traps follow:

- Slip joints must be accessible with a minimum access panel of 12 × 12 in.
- Trap seals must be 2–4 in.
- Floor drains subject to evaporation require *trap-primers* or other approved devices.
- "S" traps are prohibited.
- Vertical distance from the fixture outlet to the trap weir cannot be more than 24 in.
- Horizontal distance from the fixture outlet to the trap cannot be more than 30 in.

Plumbing Fixtures

Fixtures, faucets, and fixture fittings must have smooth, impervious surfaces and comply with the applicable referenced standards.

Fixture Clearance

Water closets, lavatories, and bidets must meet the following minimum clearance requirements:

- 15 in. from centerline of the fixture to walls or vanities

Table 14.11 Size of traps and trap arms for plumbing fixtures

Plumbing fixture	Trap size minimum (in.)
Bathtub (with or without shower head and/or whirlpool attachments)	1½
Bidet	1¼
Clothes washer standpipe	2
Dishwasher (on separate trap)	1½
Floor drain	2
Kitchen sink (one or two traps, with or without dishwasher and garbage grinder)	1½
Laundry tub (one or more compartments)	1½
Lavatory	1¼
Shower (based on the total flow rate through showerheads and body sprays) Flow rate:	
5.7 gpm* and less	1½
More than 5.7 gpm up to 12.3 gpm	2
More than 12.3 gpm up to 25.8 gpm	3
More than 25.8 gpm up to 55.6 gpm	4

*gallons per minute

- 30 in. from centerline to centerline of adjacent fixtures
- 21 in. clearance in front of fixtures

Laundry Standpipes
Standpipes must extend 18–42 in. above the trap weir.

Dishwashers

The IRC sets minimum connection requirements for dishwasher drains to prevent contamination of the dishwasher contents.

- A ¾ in. dishwasher drain may be connected to the sink tailpiece with a *wye fitting* (fig. 14.6).
- The sink, food-waste disposer, and dishwasher may discharge through a single 1½ in. trap.
- The dishwasher waste line must be looped to the underside of the counter as backflow protection.

Showers

The requirements for shower dimensions, access and scald prevention are as follows:

- Shower compartments must be at least 30 × 30 in.
- A minimum width of 25 in. is permitted if the shower area is at least 1,300 sq. in.
- Hinged shower doors must open outward.
- Finished width of at least 22 in. is required for access to shower.
- Shower valves require a water temperature high-limit stop set to no more than 120°F.

Bathtubs and Whirlpool Bathtubs

Bathtubs and whirlpool baths require the following:

- A device to limit water temperature to no more than 120°F

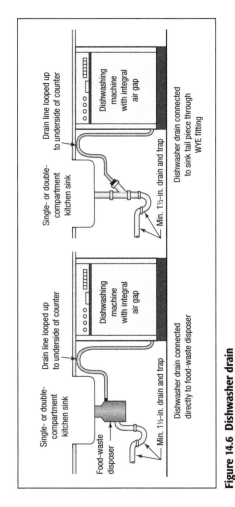

Figure 14.6 Dishwasher drain

Source: Residential Code Essentials, 2015. Washington, DC: International Code Council, 2015, p. 214.

- Access to a circulation pump as follows:
  - Sized and located in accordance with the manufacturer's installation instructions
  - At least 12 × 12 in. or at least 18 × 18 in. if the pump is more than 2 ft. from the access opening
  - An unobstructed access opening large enough to remove the pump

Water Heaters

The IRC prescribes location, installation, and safety requirements for water heaters, including drain pan installations and temperature relief valve provisions.

Required Pan for Water Heaters

If a leaking water heater will cause damage, the water heater must have a pan that meets the following minimum requirements:

- 0.0236 in. galvanized steel, aluminum, or other approved materials
- 1½ in. deep
- ¾ in. indirect waste pipe to approved waste receptor

Note: Safety pan requirements apply to storage-type water heaters, not tankless water heaters.

Pressure and Temperature Relief Valves

Pressure- and temperature-relief valves must

- Not be connected directly to the drainage system
- Discharge to one of the following at a readily observable location:
  - Floor
  - Water heater pan
  - Waste receptor
  - Outdoors
- Terminate not more than 6 in. and not less than two pipe diameters above the floor or approved waste receptor
- Not have valves, tees, traps, or a threaded end

Note: *Where PEX or PE-RT tubing is used for discharge piping, the code requires one nominal size larger than the relief-valve outlet with the end of the tubing fastened in place.*

Water Heaters Installed in Garages

When a water heater is in a garage, its ignition source must be elevated at least 18 in. above the garage floor (fig. 14.7).

Note: *The code does not require elevation of appliances that are listed as flammable vapor ignition-resistant.*

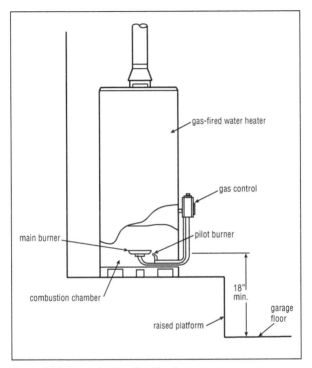

Figure 14.7 Water heater elevation in garage

Source: 2006 International Residential Code Study Companion. Washington, DC: International Code Council, 2006, p. 364.

Water Heater Seismic Bracing

In SDCs D_0, D_1, and D_2, and for townhomes located in SDC C, water heaters require bracing to resist lateral forces. The code requires anchors or straps in the upper $\frac{1}{3}$ and lower $\frac{1}{3}$ of the appliance.

15

Electrical Systems

The electrical provisions of the IRC are based on the *2014 National Electrical Code* (NEC) (NFPA 70-2014),[15] published by NFPA. Like the IRC, this chapter specifically covers typical electrical installations in constructing one- and two-family dwellings, including

- Services
- Power distribution systems
- Fixtures
- Appliances
- Devices
- Appurtenances

Methods and materials specified in the NEC (NFPA 70) are permitted, even if they are not included in the IRC. However, electrical systems or components not specifically covered in the IRC must comply with the applicable provisions of the NEC (NFPA 70).

Scope

The scope of the IRC covers single-phase services that are not more than 120/240-volt and 400 *amps*.

Equipment Location and Clearances

Following are the minimum requirements for service equipment (i.e., service panel or subpanel):

- Working space in front of equipment is at least 30 in. wide × 36 in. deep.
- Headroom for working spaces for service equipment and panelboards is at least 6 ft. 6 in.
- Illumination is provided.
- Working space must not be storage space.
- Panelboards must not be in clothes closets or bathrooms.
- Dedicated space above and below panels equals the width of panel from floor to structural ceiling (fig. 15.1).

Minimum Size of Conductors

The minimum sizes of conductors for feeders and branch circuits are 14 *AWG* copper and 12 AWG aluminum.

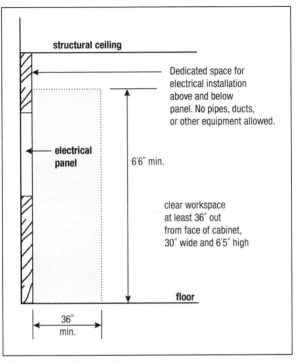

structural ceiling

Dedicated space for
electrical installation
above and below
panel. No pipes, ducts,
or other equipment allowed.

**electrical
panel**

6'6" min.

clear workspace
at least 36" out
from face of cabinet,
30" wide and 6'5" high

floor

36"
min.

Figure 15.1 Service working space and clearances

Conductor and Terminal Identification

Device terminals other than panelboards and certain devices rated over 30 amps require properly identified terminals.

Grounded (Neutral) Conductors and Terminals

The requirements for identifying *grounded conductors* and terminals are as follows:

- Grounded conductors not more than 6 AWG are identified by white or gray insulation, or bear three white stripes on other than green insulated wire.
- Grounded conductors greater than 6 AWG are white or gray, or bear three white stripes on other than green insulation, or have white or gray tape applied to terminal ends at installation.
- Receptacles and polarized attachment plugs require the grounded terminal to be identified in one of three ways as follows:
  - Substantially white in color
  - Bear the word "white"
  - Bear the letter "W"

Grounding Conductors

Identification of *grounding conductors* must comply with the following:

- Grounding conductors not more than 6 AWG are green or green with yellow stripes when not bare wire.
- Grounding conductors greater than 6 AWG have green insulation or yellow stripes on green insulation, or are bare wires. If the insulation is another color, the electrician may mark the conductors with green tape or permanent marking (or strip insulation to bare wire) at terminal ends and access points at the time of installation.

Ungrounded (Hot) Conductors

Ungrounded conductors must be identified as follows:

- Ungrounded conductors are any color other than white, gray, or green.
- White, gray, or green insulated conductors in cable or flexible cord assemblies or switch loops may be used as ungrounded conductors if they are re-identified with permanent markings.

Electrical Services

The electrical service is typically a panelboard with a main disconnect that shuts off all power to the building, the circuit breakers serving the branch circuits, and related equipment. The serving utility company delivers electrical power to the service entrance conductors, which in turn conduct the energy to the main service disconnect. The service distributes electricity to the premises wiring system.

- Only one service is permitted for one- and two-family dwellings.
- Service conductors supplying a building are not permitted to pass through the interior of another building.
- Only service conductors are permitted in the same service raceway or service cable.
- A service disconnect must be installed at a readily accessible location either outside of a building or inside, near to where the service conductors enter the building.
- Service disconnects are not permitted in bathrooms.
- Occupants must have access to the disconnect that serves their dwelling unit.

Electrical Service Size and Rating

The minimum service rating for single-family dwellings is 100 amps (60 amps for other installations). The *ampacity* of ungrounded service conductors and the rating of the service disconnect may not be less than the load served. Table 15.1 contains the formulas to calculate the minimum service load for a single-family dwelling.

Table 15.1 Minimum service load calculation for single-family dwellings

Loads and procedure				Volt-amperes
1. General lighting and general use receptacle outlets	3 volt-amperes	×	Floor area (sq. ft.)	_____ VA
				Plus
2. All 20-ampere-rated small appliance and laundry circuits	1,500 volt-amperes	×	Number of circuits	_____ VA
				Plus
3. Appliances	The nameplate volt-ampere rating of all permanent or dedicated appliances:			
	Ranges			_____ VA
	Ovens			_____ VA
	Cooking units			_____ VA
	Clothes dryers			_____ VA
	Water heaters			_____ VA
	Subtotal			**_____ VA**

(continued)

Table 15.1 Minimum service load calculation for single-family dwellings *(continued)*

Loads and procedure				Volt-amperes
4. Apply the following demand factors to the above subtotal:	100%	×	First 10,000 volt-amperes	**Plus** ____ VA
	40%	×	Portion in excess of 10,000 volt-amperes	**Plus** ____ VA
5. Air-conditioning	Total of the nameplate rating(s) of the air-conditioning and cooling equipment			**Plus** ____ VA
			Total volt-amperes	____ **VA**
Total load in amperes	**Volt-ampere sum**	÷	**240 volts** =	____ **AMPS**

Service Conductor Size

The IRC requires the following sizing of service conductors:

- Ungrounded (hot phase leg) service conductors must have a minimum size as identified in table 15.2.
- The grounded (neutral) conductor ampacity must be at least the maximum unbalance of the load. Its size must be at least the required minimum *grounding electrode conductor* size.
- The grounding electrode conductors are sized based on the size of the service entrance conductors (table 15.2).

Overhead Service-Drop and Service-Conductor Installation

The code prescribes the following minimum clearance requirements for overhead services:

- Open conductors require at least a 3 ft. horizontal separation from openings and porches and at least 3 ft. below openable windows.
- Vertical clearance height is as follows:
  - At least 3 ft. above roofs with a pitch of at least 4:12
  - At least 8 ft. above roofs with a pitch of less than 4:12
  - At least 18 in. over 4 ft. of roof overhang

Table 15.2 Service conductor and grounding electrode

Conductor types and sizes—THHN, THHW, THW, THWN, USE, XHHW, THW-2, THWN-2, XHHW-2, SE, USE-2		Allowable ampacity	Minimum grounding electrode conductor size	
Copper (AWG)	Aluminum and copper-clad aluminum (AWG)	Maximum load (amps)	Copper (AWG)	Aluminum (AWG)
4	2	100	8	6
3	1	110	8	6
2	1/0	125	8	6
1	2/0	150	6	4
1/0	3/0	175	6	4
2/0	4/0 or two sets of 1/0	200	4	2
3/0	250 kcmil or two sets of 2/0	225	4	2
4/0 or two sets of 1/0	300 kcmil or two sets of 3/0	250	2	1/0
250 kcmil or two sets of 2/0	350 kcmil or two sets of 4/0	300	2	1/0
350 kcmil or two sets of 3/0	500 kcmil or two sets of 250 kcmil	350	2	1/0
400 kcmil or two sets of 4/0	600 kcmil or two sets of 300 kcmil	400	1/0	3/0

Note: Service conductors in parallel sets of 1/0 and larger are permitted in either a single raceway or in separate raceways. Grounding electrode conductors of size 8 AWG require protection with conduit. Grounding electrode conductors of size 6 AWG require protection with conduit or must closely follow a structural surface for physical protection.

- At least 10 ft. over grade, sidewalks, and walking surfaces at service entrance point and below drip loop
- At least 12 ft. over driveways and residential property

System Grounding and Bonding

The grounding system provides a fault current path to earth. The code provides for bonding of metal parts that have potential to become energized to provide an effective path for fault current.

- System (including service equipment enclosures, *equipment grounding conductors,* and grounded service conductor) is grounded at service with grounding electrode conductor connected to grounding electrode system.
- A main *bonding jumper* (such as the green machine screw or green insulated conductor supplied with the service enclosure) is required to connect the equipment grounding conductors to the service equipment enclosure and the grounded (neutral) conductors.

Grounding Electrode System

The following electrodes, if available, are bonded together to form the grounding electrode system:

- ▦ Underground metal water pipe with at least 10 ft. in contact with the ground (must be supplemented by another type of electrode)
- ▦ 20 ft. of ½ in. reinforcing bar or 20 ft. of 4 AWG bare copper wire encased in concrete footing/foundation that is in contact with the ground
- ▦ Ground ring at least 30 in. below ground and at least 20 ft. of 2 AWG bare copper wire
- ▦ Plate electrode at least 2 sq. ft. and at least 30 in. below ground
- ▦ Listed copper clad ground rod, ⅝ in. diameter × 8 ft. long (or other approved pipe or rod electrode)
  Note: *Where ground rods are installed, at least two are required unless resistance testing indicates otherwise. A minimum separation distance of 6 ft. is required.*

Bonding

The code provides for the bonding of metal parts associated with the electrical system to provide an effective path for fault current.

- ▦ *Bonding* is required to ensure electrical continuity of the grounding system at service equipment, raceways, and enclosures.
- ▦ Bonding of the metal water piping system to the service equipment enclosure (or other approved location) with an appropriately sized bonding conductor is required to provide a fault path, in

case the water pipe becomes energized through accidental contact with hot wires. The water piping system is not a ground or *grounding electrode*. The points of attachment of the bonding jumper must be accessible.

■ For bonding of other systems—typically telephone, satellite and cable television systems—to the building grounding system, the code requires installation of an intersystem bonding termination near the building service equipment with a capacity for at least 3 bonding conductors.

Grounding Electrode Conductors

Grounding electrode conductors connect the electrical grounding system to the grounding electrodes. The following requirements apply:

■ Grounding electrode conductors may not be spliced except by an approved irreversible process.

■ Grounding electrode conductors require protection (such as approved conduit) if exposed to potential physical damage.

■ Connections to electrodes require approved fittings to ensure an effective grounding path.

Branch Circuit Ratings

Branch circuits must be rated according to the maximum allowable ampere rating or setting of the *overcurrent protection device* (table 15.3). The ampere rating or setting of the specified overcurrent device determines the circuit rating for conductors of higher ampacity.

Required Branch Circuits

The IRC requires the following branch circuits:

- Central heating equipment requires an individual branch circuit.
- Kitchen countertop receptacles require at least two, 20 amp-rated branch circuits. These circuits also are permitted to supply other receptacle outlets in the kitchen, pantry, breakfast, and dining areas, including the outlet for the refrigerator.

Table 15.3 Branch-circuit requirement

	Circuit rating		
	15 amp	**20 amp**	**30 amp**
Conductors: Minimum size (AWG) circuit conductors (copper)	14	12	10
Overcurrent-protection device: maximum ampere rating	15	20	30
Outlet receptacle rating (amperes)	15 maximum	15 or 20	30
Maximum load (amperes)	15	20	30

- The refrigerator receptacle outlet is permitted on an individual branch circuit rated 15 amps.
- The code requires a separate 20-amp laundry circuit to serve all laundry area receptacles.
- Bathroom receptacle outlet(s) require at least one 20 amp circuit that serves no other outlets.
  - The code permits the circuit to supply other equipment in the bathroom, such as a light or fan, if the circuit is dedicated to one bathroom.

Conductor Sizing and Overcurrent Protection

Table 15.4 lists ampacities for conductors. Note the following:

- The temperature rating of the conductor insulation must be suitable for the device or location.
- When conductors are bundled or stacked or there are more than three current-carrying conductors in a cable or raceway greater than 24 in. long, ampacity ratings are reduced (table 15.5).
- The temperature rating of the terminal must be considered in calculating the ampacity of the conductor.

Overcurrent Protection Required

An overcurrent device, such as a circuit breaker, must protect all ungrounded branch-circuit and feeder conductors. Overcurrent protective device

Table 15.4 Allowable ampacities

Conductor size	Conductor temperature rating		
	60°C	75°C	90°C
	Types	Types	Types
	TW, UF	RHW, THHW, THW, THWN, USE, XHHW	RHW-2, THHN, THHW, THW-2, THWN-2, XHHW, XHHW-2, USE-2
AWG kcmil	Copper		
18	—	—	14
16	—	—	18
14	20	20	25
12	25	25	30
10	30	35	40
8	40	50	55
6	55	65	75
4	70	85	95
3	85	100	110
2	95	115	130
1	110	130	150
1/0	125	150	170
2/0	145	175	195
3/0	165	200	225
4/0	195	230	260

Table 15.4 Allowable ampacities (*continued*)

Conductor size	Conductor temperature rating		
	60°C	75°C	90°C
	Types	Types	Types
AWG kcmil	TW, UF	RHW, THHW, THW, THWN, USE, XHHW	RHW-2, THHN, THHW, THW-2, THWN-2, XHHW, XHHW-2, USE-2
	Aluminum or copper-clad aluminum		
18	–	–	–
16	–	–	–
14	–	–	–
12	20	20	25
10	25	30	35
8	30	40	45
6	40	50	60
4	55	65	75
3	65	75	85
2	75	90	100
1	85	100	115
1/0	100	120	135
2/0	115	135	150
3/0	130	155	175
4/0	150	180	205

Table 15.5 Conductor proximity adjustment factors

Number of current-carrying conductors in cable or raceway	% of values in allowable ampacities table
4–6	80
7–9	70
10–20	50
21–30	45
31–40	40
41 and above	35

ratings or settings may not exceed the allowable ampacity of the conductor (table 15.6).

Location of Overcurrent Devices
Overcurrent devices must be
- Readily accessible
- Not subject to damage
- Not in clothes closets or bathrooms
- Not located over steps
- Not more than 6 ft. 7 in. above floor

Wiring Methods

The IRC recognizes a number of approved wiring methods, including armored cable, metal-clad cable,

Table 15.6 Overcurrent protection rating

Copper		Aluminum or copper-clad aluminum	
Size (AWG)	Maximum overcurrent protection device rating (amps)	Size (AWG)	Maximum overcurrent protection device rating (amps)
14	15	12	15
12	20	10	25
10	30	8	30

Note: The maximum overcurrent protection device rating shall not exceed the conductor's allowable ampacity after applying any correction or adjustment factors.

and conductors installed in various types of metallic and nonmetallic conduit. Unless stated otherwise, this section generally refers to aboveground wiring methods using nonmetallic (type NM) sheathed cable common to one- and two-family dwelling and townhome construction (table 15.7). The cable must be approved for the location. For example, type NM cable is not permitted underground, may not be used in wet or damp locations, and is not to be embedded in concrete. Other wiring methods are permitted in accordance with the code.

- Cables in accessible attics with permanent stairs require protection when installed as follows:

Table 15.7 Installation requirements for nonmetallic sheathed cable

Installation	Physical protection	
	Minimum setback from edge of framing (in.)	**Physical protection if minimum distance is not met**
Cable run parallel with the framing member or furring strip	1¼ in.	0.0625-in. steel plate or sleeve
Cable in bored holes or notches in framing members	1¼ in.	0.0625-in. steel plate or sleeve
Cable installed in grooves and covered	1¼ in. free space	0.0625 in. steel plate or sleeve
Securely fastened bushings or grommets are required to protect cable run through openings in metal framing members	—	—
	Support of cable	
Maximum allowable on center support spacing	4.5 ft.	
Maximum support distance from metal box with cable clamp	12 in.	
Maximum support distance from plastic box without cable clamp	8 in.	
Flat cables shall not be stapled on edge	—	

- Across the tops of joists
- On the faces of studs within 7 ft. above the floor joists
- On the faces of rafters within 7 ft. above the floor joists

▦ Cables in accessible attics without permanent stairs must be protected only if they are within 6 ft. of the attic entrance.

▦ Exposed cable must
  - Closely follow the surface of the building finish or running boards
  - Be protected from physical damage
  - Have a sunlight-resistant listing if exposed to sunlight

▦ Unfinished basements with type SE or NM cable that runs at angles with joists must adhere to the following requirements:
  - Cables with two or more conductors at least 6 AWG may be attached directly to the bottom of joists.
  - Cables with three or more conductors at least 8 AWG may be attached directly to the bottom of joists.
  - Smaller cables must run either through bored holes in joists or on running boards.

Note: *For wall installations in unfinished basements, Type NM cable installed in a listed conduit or tubing satisfies the physical protection requirement.*

Underground Installation Requirements

Cable installed underground must be listed and labeled for the location. In addition, direct burial conductors and cables emerging from the ground must be protected from 18 in. below grade to 8 ft. above grade (table 15.8). Direct burial conductors emerging from the ground and subject to physical damage require one of the following for protection:

- Rigid metal conduit
- Intermediate metal conduit
- Schedule 80 PVC conduit

Note: *Splices underground are permitted only with specific approved methods and listed materials.*

General Purpose Receptacle Distribution

In addition to providing convenience for the occupants, proper placement of receptacle outlets to serve fixtures and appliances can reduce electrical hazards in the home that may occur with the use of extension cords. The IRC provides minimum distribution requirements for general purpose receptacle outlets rated at 125 volts, 15 and 20 amps.

Habitable Rooms

Any point along a wall must be within 6 ft. of a receptacle outlet (fig. 15.2).

Kitchen Counters

The location requirements for receptacle outlets serving kitchen counters are as follows:

- Any point along a kitchen wall counter must be within 2 ft. of a receptacle outlet (fig. 15.3).
- Receptacle outlets must serve island and peninsular counter spaces 24 in. long and 12 in. wide or larger.
- Required receptacles may be no higher than 20 in. above counter.
- Receptacles are permitted 12 in. or less below peninsular and island counters.

Bathroom

A receptacle outlet is required within 36 in. of each lavatory.

Outdoors

A receptacle outlet no more than 6 ft. 6 in. above grade or walking surface is required at each of the following locations:

- Front of dwelling unit
- Back of dwelling unit
- Each balcony, deck, or porch that is accessed from inside the dwelling unit

Table 15.8 Minimum cover requirements, burial (in.)

Location of wiring method or circuit	Type of wiring method or circuit	
	Direct burial cables or conductors	Rigid metal conduit or intermediate metal conduit
All locations not specified below	24	6
In trench below 2-in.-thick concrete or equivalent	18	6
Under a building	0 (in raceway only)	0
Under minimum of 4-in.- thick concrete exterior slab with no vehicular traffic and the slab extending not less than 6 in. beyond the underground installation	18	4
Under streets, highways, roads, alleys, driveways and parking lots	24	24
One- and two-family dwelling driveways and outdoor parking areas, and used only for dwelling-related purposes	18	18
In solid rock where covered by minimum of 2 in. concrete extending down to rock	2 (in raceway only)	2

Table 15.8 Minimum cover requirements, burial (in.) (*continued*)

	Type of wiring method or circuit	
Nonmetallic raceways listed for direct burial without concrete encasement or other approved raceways	**Residential branch circuits rated 120 volts or less with GFCI protection and maximum overcurrent protection of 20 amperes**	**Circuits for control of irrigation and landscape lighting limited to not more than 30 volts and installed with type UF or in other identified cable or raceway**
18	12	6
12	6	6
0	0 (in raceway only)	0 (in raceway only)
4	6 (direct burial) 4 (in raceway)	6 (direct burial) 4 (in raceway)
24	24	24
18	12	18
2	2 (in raceway only)	2 (in raceway only)

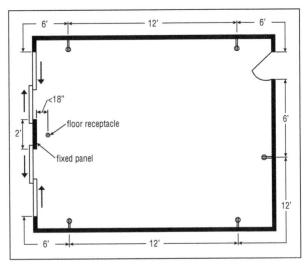

Figure 15.2 Receptacle outlet locations

Source: 2006 International Residential Code Study Companion. Washington, DC: International Code Council, 2006, p. 438.

Laundry Areas
At least one outlet is required for laundry appliances.

Basements, Garages, and Accessory Buildings
Basements, garages, and accessory buildings require receptacle outlets as follows:
- Each separate unfinished portion of a basement requires at least one receptacle outlet.

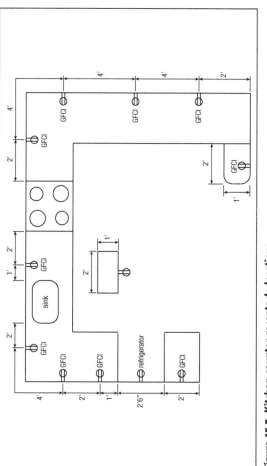

Figure 15.3 Kitchen counter receptacle locations

Source: 2006 International Residential Code Study Companion. Washington, DC: International Code Council, 2006, p. 439.

- Attached garages and detached garages with electrical power require at least one receptacle outlet for each vehicle space.
- Detached accessory buildings with electrical power require at least one receptacle outlet.

Hallways
The code requires at least one receptacle outlet in hallways 10 ft. or longer.

Foyers
For foyers greater than 60 sq. ft., the code requires a receptacle outlet for each wall space 3 ft. or greater in width.

HVAC Equipment
A receptacle outlet is required within 25 ft. of HVAC equipment for service of the equipment.

Ground-Fault Circuit Interrupter (GFCI) Protection

Single-phase 125-volt receptacle outlets of 15 and 20 amps at the following locations require *GFCI* protection:
- Bathrooms
- Garages and accessory buildings
- Outdoors

- Crawl spaces
- Unfinished basements
- Kitchen receptacles serving countertops
- Within 6 ft. of sinks, including laundry, utility, and bar sinks
- Within 6 feet of the outside edge of a bathtub or shower stall
- Laundry areas
- Kitchen dishwasher branch circuit

Note: *Receptacle outlets located in unfinished basements and serving permanently installed burglar or fire alarm systems do not require GFCI protection.*

Note: *GFCI devices must be readily accessible and cannot be installed behind appliances such as refrigerators.*

Arc-Fault Circuit Interrupter (AFCI) Protection

AFCIs detect unwanted arcing in the wiring of the branch circuit and open the circuit before excessive heat buildup can cause a fire. Typically, a combination-type arc-fault circuit breaker protecting the entire branch circuit satisfies this requirement, but the code does offer other options. AFCIs are required for branch circuits supplying 120-volt, single-phase, 15- and 20-amp outlets, including receptacle, lighting and smoke alarm outlets, installed in the following locations and similar rooms or areas:

- Bedrooms
- Closets
- Dens
- Dining rooms
- Family rooms
- Hallways
- Kitchens
- Laundry areas
- Living rooms
- Libraries
- Parlors
- Recreation rooms
- Sunrooms

Lighting Outlets

Wall-switch-controlled lighting outlets are required at the following locations:

- Habitable rooms
  Note: *Switch-controlled receptacle outlets satisfy the requirement in habitable rooms, except in kitchens.*
- Bathrooms
- Stairways
  Note: *Switches must be located at each level when the stair has six or more risers.*
- Outside each exterior door with grade-level access
- Hallways

- Attached garages and detached garages with power
- Storage and equipment spaces, including attics

Boxes

A box is required at each splice, junction, outlet, switch, and pull point. Following are other requirements for boxes:

- All metal boxes require grounding by approved means.
- Type NM cable must extend into a nonmetallic box at least ¼ in. and be anchored within 8 in. of the box.
- Boxes supporting luminaires that weigh more than 50 lb. must be listed and marked for the purpose.
- Boxes that support ceiling paddle fans must be listed and marked for the purpose and, when designed to support more than 35 lb., must state the maximum weight.
- Boxes may not support ceiling paddle fans weighing more than 70 lb.
- Each box is limited to a maximum box fill capacity based on the size of the box, the number and size of conductors entering the box, and the devices and fittings installed.
- Unused openings must be closed by approved methods.

- Boxes may be set back no more than ¼ in. from noncombustible finishes such as ceramic tile and concrete.
- Boxes must be flush with or project in front of combustible finish surfaces such as wood and drywall.

Receptacles

The code requires the following for installation of receptacle devices:

- A single receptacle installed on an individual branch circuit must have an ampere rating not less than that of the branch circuit.
- Receptacles connected to a branch circuit supplying two or more receptacles require an amp rating as specified in table 15.3.
- Receptacles of 15 and 20 amps installed in a wet location, whether in use (with a cord plugged in) or not, require enclosures, such as bubble covers, that are weatherproof.
- The code prohibits receptacles within or directly over a bathtub or shower space.

Tamper-Resistant Receptacles

In general, the code requires tamper-resistant receptacles for all 125 volt, 15 and 20 amp receptacles associated with dwelling units and garages.

The following applications do not require tamper-resistant receptacles:

- Receptacles located more than 5.5 ft. above the floor.
- Receptacles that are part of a luminaire or appliance.
- A single receptacle for a single appliance or a duplex receptacle for two appliances where located in dedicated space behind the appliance.
- Receptacles in accessory buildings that are not garages
- Receptacles in attics and crawl spaces

Fixtures

The code regulates electrical fixtures for type, location, and clearance to combustibles.

Bathtub and Shower Areas

Specific requirements apply for fixture installations in a zone measured 3 ft. horizontally and 8 ft. vertically from the top of a bathtub rim or shower stall threshold as follows:

- Cord-connected or suspended luminaires, track lighting, pendants, and ceiling-suspended (paddle) fans are prohibited.
- Luminaires must be listed for damp locations.
- Luminaires subject to shower spray must be listed for wet locations.

Luminaires in Clothes Closets

To reduce fire hazard, the code restricts luminaires adjacent to storage areas of clothes closets. The storage area comprises the following space:

- Within 24 in. of the side and back walls and the greater of 6 ft. high or to the highest rod
- The space above 6 ft. that is within 12 in. (or the shelf width, if greater) of the side and back walls

Permitted Luminaires

The following types of luminaires are permitted in clothes closets:

- Surface-mounted or recessed incandescent, with completely enclosed lamps
- Surface-mounted or recessed fluorescent
- LED luminaires

Prohibited Luminaires

The following types of luminaires are prohibited in clothes closets:

- Incandescent luminaires with open or partially enclosed lamps
- Pendant luminaires
- Lamp holders

Luminaire Installation and Clearances in Clothes Closets

Luminaires typically must be installed on or recessed into the ceiling or the wall above the door. Specified clearances are measured from the fixture to the nearest point of the defined storage space (fig. 15.4). Minimum clearances are as follows:

- At least 12 in. for surface-mounted incandescent and LED luminaires
- At least 6 in. for recessed incandescent and LED luminaires with a completely enclosed lamp
- At least 6 in. for surface-mounted and recessed fluorescent luminaires

Note: *When specifically identified as suitable for installation within the closet storage space, both surface mounted LED luminaires and surface mounted fluorescent luminaires may be installed anywhere in a clothes closet in accordance with the manufacturer's installation instructions.*

Recessed Luminaire Installation and Clearance

The IRC prescribes the following clearances for recessed luminaires:

- All recessed parts must be at least ½ in. from combustible materials, unless the luminaire is identified for contact with insulation (Type IC).
- Thermal insulation cannot be installed above a recessed luminaire or within 3 in. of the recessed luminaire unless the luminaire is identified for contact with insulation (Type IC).

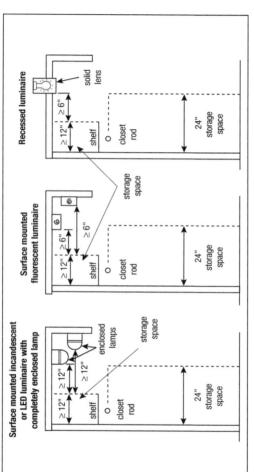

Figure 15.4 Luminaires in clothes closets

Note: Surface-mounted fluorescent or LED luminaires identified as suitable for storage space are permitted anywhere in a clothes closet when installed according to the manufacturer's instructions.

Epilogue

Communication between the builder and the local building department is essential to completing construction correctly, on schedule, and within budget. In addition to providing information on local amendments and regional design criteria, the local authority is responsible for interpreting the code and establishing administrative procedures for permits, inspections, testing, appeals, and certificates of occupancy.

Home Builders' Jobsite Codes has been developed as an accurate, abbreviated guide to the frequently used provisions of the *International Residential Code (IRC)*. To protect public health, safety, and welfare, the IRC sets minimum requirements for the construction of one- and two-family dwellings and townhomes. It is a comprehensive residential code that emphasizes prescriptive methods—specific lists of rules the home builder may follow to comply with the code.

Notes

[1] *2015 International Building Code*. Washington, DC: International Code Council, Inc., 2014.

[2] *ASCE/SEI 7 (2010) Minimum Design Loads for Buildings and Other Structures with Supplement No. 1*. Reston, VA: American Society of Civil Engineers, Structural Engineering Institute, 2010.

[3] *ICC 600 (2014) Standard for Residential Construction in High-wind Regions*. Washington, DC: International Code Council, Inc., 2014.

[4] *Wood Frame Construction Manual for One- and Two-family Dwellings (AWC WFCM—2015)*. Leesburg, VA: American Wood Council, 2015.

[5] *ICC 500 (2014) Standard on the Design and Construction of Storm Shelters*. Washington, DC: International Code Council, Inc., 2014.

[6] *NFPA 13 (2013) Standard for the Installation of Sprinkler Systems*. Quincy, MA: National Fire Protection Association, 2012.

[7] *NFPA 13R (2013) Standard for the Installation of Sprinkler Systems in Residential Occupancies Up to and Including Four Stories in Height.* Quincy, MA: National Fire Protection Association, 2012.

[8] *NFPA 13D (2013) Standard for the Installation of Sprinkler Systems in One- and Two-family Dwellings and Manufactured Homes.* Quincy, MA: National Fire Protection Association, 2012.

[9] *2015 International Energy Conservation Code.* Washington, DC: International Code Council, Inc., 2014.

[10] *2015 International Mechanical Code.* Washington, DC: International Code Council, Inc., 2014.

[11] *2015 International Fuel Gas Code.* Washington, DC: International Code Council, Inc., 2014.

[12] *NFPA 31 (2011) Standard for the Installation of Oil-burning Equipment.* Quincy, MA.: National Fire Protection Association, 2011.

[13] *2015 International Fuel Gas Code.* Washington, DC: International Code Council, Inc., 2014.

[14] *2015 International Plumbing Code.* Washington, DC: International Code Council, Inc., 2014.

[15] *NFPA 70 (2014) National Electrical Code.* Quincy, MA.: National Fire Protection Association, 2013.

Glossary

accessory structure. A building, such as a detached garage or shed, limited to 3,000 sq. ft. and two stories in height, and located on the same lot as the dwelling.

AFCI. Arc-fault circuit interrupter, a device that detects unwanted arcing in the wiring of the branch circuit and opens the circuit before excessive heat buildup can cause a fire.

air gap. Open space between the potable water outlet and the flood-level rim of the fixture, designed to separate the potable water from the source of contamination.

air-impermeable insulation. An insulation having an air permeance equal to or less than 0.02 L/s-m^2 at 75 Pa pressure differential when tested according to ASTM E 2178 or E 283.

ampacity. The current, in amperes, that a conductor can carry continuously under the conditions of use without exceeding its temperature rating.

amps. Amperes, a measurement of electrical current.

attic. The unfinished space between the ceiling assembly of the top story and the roof assembly

AWG. American Wire Gage. The term indicates the size of the wire. As the AWG number decreases, the wire diameter increases.

bathroom. A room containing a bathtub, shower, spa, or similar bathing fixture.

bonding. Permanent joining of metallic parts to form an electrically conductive path that will ensure electrical continuity and the capacity to conduct safely any current likely to be imposed. Bonding is not the same as grounding, although the two terms are sometimes confused.

bonding jumper. A reliable conductor to ensure the required electrical conductivity between metal parts required to be electrically connected

braced wall line. A straight line through the building plan that represents the location of the lateral resistance provided by the wall bracing

braced wall panel. A full-height section of wall constructed to resist in-plane shear loads through interaction of framing members, sheathing material, and anchors. The panel's length meets the requirements of its particular bracing method, and contributes toward the total amount of bracing required along its *braced wall line*.

Btu/h. British thermal units per hour

building drain. Lowest drainage piping inside the house; extends 30 in. beyond the exterior walls to connect to the building sewer. The building drain collects the discharge from all other drainage piping in the dwelling.

building sewer. That part of the drainage system that begins 30 in. outside of the building, at the end of the building drain, and conveys its discharge to a public sewer, private sewer, individual sewage disposal system, or other approved point of disposal.

Category IV appliance. An appliance that operates with a positive vent static pressure and with a vent gas temperature that is capable of causing excessive condensate production in the vent. Category IV appliances are referred to as very high efficiency condensing appliances. They rely on mechanical means rather than gravity to vent the low temperature flue gases to the outside. Vents must be noncorrosive material, such as PVC pipe, or as specified by the appliance manufacturer.

cfm. Cubic feet per minute

cricket. A sloped flashing on the up-roof side of a chimney to divert water from above the chimney to each side

dead loads. Weight of all materials of construction incorporated into the building, including but not limited to walls, floors, roofs, ceilings, stairways, built-in partitions, finishes, cladding, and fixed-service equipment. Dead loads are permanent.

d.f.u. In plumbing, drainage fixture units, a measurement of probable discharge into the drainage system by various types of plumbing fixtures, used to size DWV piping systems.

DWV. In plumbing, the drainage, waste, and vent system. It consists of all pipes that convey wastes from plumbing fixtures, appliances, and appurtenances, including fixture traps, above-grade drainage piping, below-grade drains within the building, below- and above-grade venting systems, and piping to the public sewer or private septic system.

direct-vent appliance. Fuel-burning appliance with a sealed combustion system that draws all air for combustion from the outside atmosphere and discharges all flue gases to the outside atmosphere

dwelling. A building that contains one or two dwelling units occupied or intended to be occupied for living purposes

dwelling unit. A single unit providing complete independent living facilities for one or more persons, including permanent provisions for living, sleeping, eating, cooking, and sanitation.

equipment grounding conductor. Conductor used to connect the noncurrent-carrying metal parts of equipment, conduit, and other enclosures to the system-grounded (neutral) conductor, the grounding electrode conductor, or both, at the service equipment.

fenestration. Skylights, roof windows, vertical windows (whether fixed or moveable); opaque doors; glazed doors; glass block; and combination opaque/glazed doors.

flashover. The sudden and rapid spread of fire caused by the near-simultaneous ignition of all combustible smoke, fumes, and exposed materials in an enclosed space.

flight. A continuous run of stair treads between landings.

footcandle. A unit of measure of the intensity of light falling on a surface, equal to one lumen per square foot and originally defined with reference to a standardized candle burning at one foot from a given surface.

GFCI. Ground-fault circuit interrupter. A device intended to protect people, it functions to de-energize a circuit within an established period of time when a current to ground exceeds a certain value (indicating a fault).

grade plane. A reference plane representing the average of the finished ground level adjoining the building at all exterior walls. Where the finished ground level slopes away from the exterior walls, the grade plane is established by the lowest points within the area between the building and the lot line. Where the lot line is more than 6 ft. from the building the grade plane is established by the lowest points between the structure and a point 6 ft. from the building.

gray water. Waste discharged from lavatories, bathtubs, showers, clothes washers and laundry trays.

grounded conductor. System or circuit conductor that is intentionally grounded. This is commonly referred to as the *neutral* and is sometimes confused with the terminology for grounding conductor. Grounded conductors (wires) generally are marked with white.

grounding conductor. A conductor used to connect equipment, or the grounded circuit of a wiring system to a grounding electrode or electrodes. The grounding conductor typically bonds to the grounded (neutral) conductor at the service equipment only and typically has green markings or is bare wire.

grounding electrode. Device that establishes an electrical connection to earth. Ground rods and underground copper water service piping are examples of grounding electrodes.

grounding electrode conductor. Conductor used to connect the grounding electrode(s) to the equipment grounding conductor, to the grounded conductor, or to both, typically at the service.

habitable attic. A finished or unfinished area, not considered a story, with an occupiable floor area at least 70 sq. ft. complying with the ceiling height requirements and enclosed by the roof assembly above and the floor-ceiling assembly below.

habitable space. Space in a building for living, sleeping, eating, or cooking. Bathrooms, toilet rooms, closets, halls, storage or utility spaces, and similar areas are not considered habitable spaces.

high-efficacy lamps. Compact fluorescent lamps, T-8 or smaller diameter linear fluorescent lamps, or lamps with a minimum efficacy of 60 lumens per watt for lamps over 40 watts; 50 lumens per watt for lamps of more than 15 watts to 40 watts; and 40 lumens per watt for lamps of 15 watts or less.

interlayment. In the application of wood shakes, interlayment is an 18-inch-wide strip of at least No. 30 felt shingled between each course of shakes so that no felt is exposed to the weather.

jalousie. A window, blind, or shutter with adjustable horizontal slats. A jalousie window has overlapping glass slats that open to allow the passage of air and light.

labeled. Equipment, materials, or products bearing a label, seal, symbol or other identifying mark of a nationally recognized testing laboratory, inspection agency, or other organization whose labeling indicates either that the equipment, material, or product meets identified standards or that it has been tested and found suitable for a specified purpose.

listed. Equipment, materials, products, or services included in a list published by an approved organization that states either that the equipment, material, product, or service meets identified standards or that it has been tested and found suitable for a specified purpose.

live load. Variable loads produced by the use and occupancy of the building, such as people and furnishings, not including construction or environmental loads produced by dead loads or by wind, snow, rain, earthquake, flood, or dead loads.

local exhaust. An exhaust system that uses one or more fans to exhaust air from a specific room or rooms (typically bathrooms and kitchens) within a dwelling.

luminaire. Complete lighting unit (lighting fixture) consisting of a lamp or lamps, together with parts designed to distribute the light, to position and protect the lamps and ballast, and to connect the lamps to the power supply.

makeup air. Air that replaces the air being exhausted from a room or structure

means of egress. A continuous and unobstructed vertical and horizontal path from any occupied portion of a dwelling to the exterior of the dwelling at the required egress door, without requiring travel through a garage.

O.C. On-center spacing of studs, joists, rafters, or fasteners.

on-site non-potable water reuse systems. Water systems for the collection, treatment, storage, distribution, and reuse of non-potable water generated on site, including but not limited to graywater systems. This definition does not include rainwater harvesting systems.

overcurrent protection device. A circuit breaker, fuse, or other device that protects the circuit by opening the device, thereby disconnecting power to the circuit, when the current reaches a value that causes excessive heat in conductors (overcurrent).

performance. Design and construction to ensure the building system will function in a certain way to meet the building code's minimum requirements. Performance to meet structural requirements is typically achieved through engineering.

prescriptive. When referring to code provisions, a specific set of rules the builder may follow to assure code compliance.

psf. Pounds per square foot

PVC. Polyvinyl chloride

R-value. A measure of resistance to heat flow through a given thickness of material, typically insulation. A higher R-value means a greater thermal resistance.

rabbeted. With a recess or groove cut into the edge, as in a piece of wood.

racking. Movement of structural elements out of level or plumb by the lateral forces of wind, earthquake ground motion, or other loads.

safe room. A storm shelter

seismic. Characteristic of or having to do with earthquake ground motion

seismic design category (SDC). A classification assigned to a structure based on its occupancy category and the severity of the design earthquake ground motion at the site

solar heat gain coefficient (SHGC). The ratio of the solar heat gain entering the space through a window assembly to the incident solar radiation. The lower a window's solar heat gain coefficient, the less solar heat it transmits.

stack vent. Continuation of the waste stack above the highest horizontal drain connection

structural composite lumber. Structural members manufactured using wood elements bonded together with exterior adhesives

structural insulated panel (SIP). A panel that consists of a lightweight foam plastic core securely laminated between two thin, rigid wood structural panel facings.

toilet room. A room containing a water closet and, frequently, a lavatory, but not a bathtub, shower, spa, or similar bathing fixture.

townhome. Same as townhouse. A dwelling unit constructed within a group of three or more attached units in which each unit extends from foundation to roof and with open space on at least two sides.

trap primer. A device that automatically replenishes water in a trap to maintain the trap seal. Trap primers typically serve floor drains that may lose their trap seal through evaporation because of infrequent use.

U-factor. The rate of heat loss through a window or door assembly. The lower the U-factor, the greater the resistance to heat flow.

ungrounded conductor. Generally referred to as the "hot" conductor (wire). The insulation color is typically black or red, but may be of any color other than white, green, or gray.

waste receptor. A floor sink, standpipe, hub drain or a floor drain that receives the discharge of one or more indirect waste pipes.

waste stack. A main line of vertical DWV piping that conveys only liquid sewage not containing fecal material

whole-house mechanical ventilation system. An exhaust system, supply system, or combination thereof, designed to mechanically exchange indoor air for outdoor air when operating continuously or through a programmed intermittent schedule to satisfy the whole-house ventilation rate.

windborne debris region. Areas within hurricane-prone regions located in accordance with one of the following:

- Within 1 mile of the coastal mean high water line where the ultimate design wind speed, V_{ult}, is 130 mph or greater.
- In areas where the ultimate design wind speed, V_{ult}, is 140 mph or greater; or Hawaii.

wood structural panel. A panel manufactured from veneers or wood strands or wafers bonded together with waterproof synthetic resins or other suitable bonding systems. Examples of wood structural panels are plywood, OSB, or composite panels.

wye fitting. Also referred to as a Y-fitting. A plumbing fitting with a Y shape, usually with one arm at a 45° angle to the main fitting.

Index